Exit Here
for Fish!

Exit Here for Fish!

Enjoying and Conserving New Jersey's Recreational Fisheries

GLENN R. PIEHLER

Rutgers University Press

NEW BRUNSWICK, NEW JERSEY,
AND LONDON

Library of Congress Cataloging-in-Publication Data
Piehler, Glenn R., 1943–
    Exit here for fish : enjoying and conserving New Jersey's recreational
fisheries / Glenn R. Piehler.
        p.   cm.
    Includes bibliographical references and index (p.).
    ISBN 0-8135-2784-8 (cloth : alk. paper) — ISBN 0-8135-2785-6
(pbk. : alk. paper)
    1. Fishes—New Jersey.   2. Fishing—New Jersey.   I. Title.

QL628.N5 P54 2000
799.1'09749—dc21   99-053440

British Cataloging-in-Publication data for this book is available
from the British Library.

Manufactured in the United States of America

To the memory of Ruth and George Piehler,
my Mom and Dad

Contents

Contents

List of Illustrations

List of Figures

List of Tables

Preface

A good many years and a lot of fish have slipped through my hands since I caught my first one, a fluke, in the New Jersey surf. Six of those years were spent studying fish in graduate school, many additional years raising two kids to be sport fishermen, and far too many working a forty-plus-hour week. Suddenly finding myself with time on my hands in March 1997, I was prompted to share some of my experience as a "fish person" and native Jersey boy. What tempted me was an article in a Sunday edition of the *New York Times* that quoted a Rutgers University Press spokesperson as saying, "I'd love to find the right person to write a guide to the area's fresh- and salt-water fishing." Being less bashful at age fifty-four than I was at seven, I called to express my interest. That fateful call and subsequently successful proposal brought me under contract by fall '97, and ultimately working countless hours—albeit a labor of love—to complete my mission. In a nutshell, that mission would be to create a popular but comprehensive account of our fresh- and saltwater sport fisheries.

I began this undertaking by defining the area's fresh- and saltwater sport fishes as those species that are, for the most part, common to New Jersey, New York, Delaware, and (with the exception of strictly coastal species) Pennsylvania. The shads pictured on the cover are emblematic of sport fishes inhabiting both fresh and salt waters during their life cycle, and whose well-being depends upon all four states.

Next I decided to interpret "guide" as more inclusive than a mere compilation of maps, directions, and fishing tips for a chosen set of fish and waterbodies, since some good books of that nature were already published. Though I had to restrict the orbit of my directory of representative fishing areas to New Jersey waters, I have tried to em-

Calvin and Hobbes

by Bill Watterson

Fig. 0.1. CALVIN AND HOBBES © 1986 by Watterson. Reprinted with permission of UNIVERSAL PRESS SYNDICATE. All rights reserved.

brace fishing opportunities from Cape May to High Point, and from the ocean and Hudson River in the east to the Delaware in the west.

I have not confined the orbit of information about fish and their habitats to basic descriptions and functional advice, opting instead to fuse together portraits of fish, fish foods and imitations, fishing anecdotes, and fish management issues with discussion of the physical and ecological characteristics of our aquatic habitats, and commentary about their origins and geological foundations. My goal was to create a piece of work that would be technically sound, well rounded, well illustrated, interesting, and "user friendly." The result draws upon my professional memoirs and interests, and no small amount of personal fishing experience here and elsewhere around the country.

In designing the book, I revisited my traditional fish biology texts, reviewed other (and perhaps newer) works concerning some subjects, and sought out publications of a regional or New Jersey–specific nature that would complement my own expertise and experience. I spent long hours pondering a logical flow, style, and balance between oversimplicity and overambitiousness in content and detail, and each step of the way found myself making adjustments in thinking or begging more information. In addition to doing more library research and sketching or photographing many of the illustrations contained herein, I tried to squeeze in as much "ground truthing" as time would permit in order to minimize bum steers relative to those places I had not previously fished, or maybe had changed.

To supplement this, I made a practice of examining weekly reports of fishing activity and issues chronicled in newspaper accounts and *The Fisherman* (an informative weekly magazine published in Point Pleasant) from 1997 through early 1999, citing such reports liberally if the shoe fit. Finally, I was astonished at the amount of information accessible via the "net" (including daily fishing reports from charter and party boat captains, bait-and-tackle shop proprietors, and avid anglers), and reference is made to web sites that may come in handy when planning a day trip. (Please note, of course, that web sites may change, come, or go.)

The payoff, I hope, is a book that will be useful to the average angler, tourist, and/or newcomer to New Jersey who might be interested in having a "one-stop" compendium of our principal sport fishes and great variety of fishing resources; those who would like to learn a bit more about the study and management of fishes and their environment, including a condensed history of the profession in America; or the high school or undergraduate college student who may be imagining a possible career in fisheries.

Illus. 1. The author, proudly displaying his "first fish," ca. 1950

New Jersey has a lot to offer to the would-be rural, suburban, or ur-bant angler. I, for one, have returned to this state following stints in Michigan, Massachusetts, New Hampshire (mid-1965 through 1973), and more recently California (1988–1994), and lived in towns typifying all three levels of development in New Jersey. With some guidance, and a good set of county maps, you can find opportunities off any Parkway, Turnpike, or Interstate exit. Naturally, as in the case of any state, it helps if you've lived there awhile, been there before, or know someone who has. For example, no one outside of Fair Haven, Red Bank, or Rumson probably knows anything about McCarter's or Schwenker's ponds. But, if you have some advance notion of what kinds of fish are likely to inhabit particular types of waterbodies around the state, you should be one step ahead of the game. That's part of the reasoning behind this book — to help avert boredom that might lead to dumping your fishing companion in the drink!

Exit Here
for Fish!

CHAPTER 1

Introduction

The pin of the arcade wheel at Seaside Heights came to rest in the slot Dad had wagered his dime on, and I wasted no time choosing a prize. I picked a five-foot conventional boat rod with a simple reel. On Dad's next spin, he won one for my younger brother, Doug. Together the next morning we picked up some sinkers, swivels, hooks, and squid, and set out on what would become a lifelong hobby and, for me, career.

It was the late forties or early fifties. Population density, distribution of homes relative to workplaces, number of cars per family, number of highways and lanes per highway, and hence wildlife habitat and public access were quite different then. The Garden State Parkway had not yet been opened. Once off Route 34 south of Matawan, having passed the always eye-catching series of Burma Shave signs, Dad's old Kaiser poked along the original Route 35 to Ocean Beach. Route 35 was a two-lane secondary road all the way to Seaside Heights until the late fifties, when it was split into north and south arteries at Metedeconk. The present southbound lanes traverse fill heaped into what had been salt marshes lining the inside of the barrier beach, thereby expanding our summer Mecca and boat-docking spaces.

I grew up in Union County, which in 1800 boasted of thriving commercial fisheries in Elizabeth and Linden, and our family bought one of the first little plywood cottages in the sand dunes of what was to become Ocean Beach. Doug and I could fish practically anywhere along the surf or in the little tidal inlets and channels of Barnegat Bay. In less than a decade that would become more difficult, but new friends and beach parties that came with the territory made for a most enjoyable and fertile time "down the shore." Many of those friends

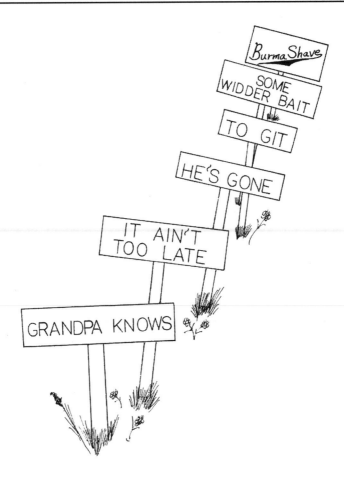

Fig. 1.1. Burma Shave jingle from Frank Rowsome's book *The Verse by the Side of the Road* as it might have appeared along Route 34 in the forties or fifties.

were also fishermen. During the off-season, our Union-based "Jersey Coasters" fishing club persisted in making excursions to Island Beach, Sandy Hook, or the Brielle and Point Pleasant party boat fisheries. Smelling the salt air and driving through towns that were, in those days, all but evacuated after Labor Day caused me to reminisce and look forward to the next summer's hubbub.

Meanwhile, the passion I had developed for freshwater fishing blossomed. My buddy Jim's dad got Doug and me started on that when we were ± ten years old. The memory of that first cold, rainy

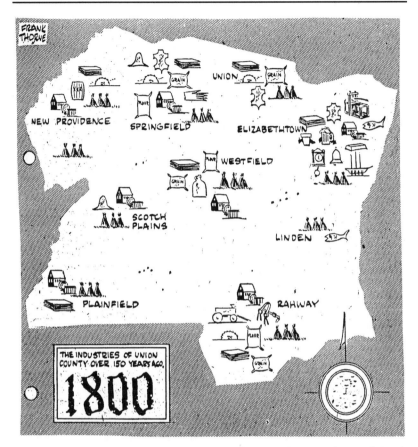

Fig. 1.2. Union County in 1800. From a February 20, 1951, issue of Frank Thorne's series entitled "The Illustrated History of Union County," which ran, based on Mom's scrapbook, from January through August in the former *Elizabeth Daily Journal*.

opening day of trout season on the Rockaway River in the middle of Dover will always be with me. The river was littered with discarded old drums (containing who knows what), tires, and engine blocks — and laced with minimally treated industrial effluent. Doug and I caught no fish that day, but we enjoyed the challenge and watching more experienced anglers land their trout. It also provided some perspective on water quality and habitat, then and now.

There were no Routes 78, 80, 280, or 287, and Routes 10, 46, and 23 were largely undeveloped beyond Livingston, Caldwell, and Pompton Plains, respectively. Public access to streams and lakes abounded, and access to more pristine or secluded habitat could be gained simply by requesting a farmer's permission to cross the pasture. That too changed as the state's population increased by a factor of almost 1.5 between 1950 and 1970 (more than 2 million people), suburban sprawl crept outward from the major metropolitan areas, and a small but conspicuous number of fishermen began abusing the privilege by trampling the landowners' plants and leaving piles of litter. Still, thanks to the joint efforts of the state, some water purveyors, and fishing and/or conservation organizations, New Jerseyans continue to enjoy a wealth of fishing opportunities.

New Jersey has over 6,000 miles of rivers and streams; 24,000 acres of public lakes, reservoirs, and ponds; 420 square miles of open bay and estuary waters; and 120 miles of ocean coast (NJDEP 1998a). Some relatively pristine areas remain despite our history of industrial, agricultural, urban, and suburban development. Over the past forty years many new waterbodies have been created, new species have been introduced, which I have yet to tackle, and the amount of toxins assaulting our inland waters — such as that first place I drowned a worm in the Rockaway River—has generally diminished with implementation of dumping restrictions and effluent limitations.

This book opens with a chapter on the taxonomic origins and classification of sport fishes I have chosen to include, and what I hope is just enough technical jargon and information on the general biology of fishes to make the remaining chapters more winning. Chapter 3 describes freshwater, and chapter 4 saltwater, sport fisheries. In total, fifty-nine species of fish are illustrated and seventy-two are discussed. Most of these can be categorized as "sport" fishes in the sense that people truly seek them, but I was compelled also to include some baitfish that ought to be recognized, as well as a number of creatures you might unwittingly hook into. In each case I have tried to capture the essence of the species or group of species — what they look like, how big they get, where they came from, what kinds of waterbodies they live in, what they do for a living, generally how and when they may be caught, how they've fared over the years and are doing today, and where you can find more specific information about some of them.

4

The second halves of chapters 3 and 4 describe the aquatic ecosystems of New Jersey. For inland (fresh, or "sweet," as the Europeans would say) waters, I have broken these ecosystems down into six categories of running (*lotic*) waterbodies and three classes of standing, or *lentic*, waterbodies. The breakdowns are a fabrication of my own reasoning (as far as I can determine), and are meant to provide the reader with a utilitarian but fundamentally not *un*scientific basis for evaluating fishing alternatives. In each case, you will find a description of what I refer to as a "type" of waterbody (or section of waterbody), with a description of that type's characteristics and what sorts of fish one might expect to catch there. Similarly, but in accord with traditional breakdowns based on physical form and variations in salinity, major categories of saltwater habitats and fisheries are described. Both fresh- and saltwater chapters are laced with facts about their geology and beginnings dating to the last great ice age.

My next chapter is entitled "Factors in Distribution and Abundance of Fishes." It is divided into subchapters on "The Importance of Habitat," "How Clean Is Clean?" and "Fishing for Fun or for a Living," and it is offered as an outgrowth of the frequency with which people ask questions about controversial subjects like wetlands preservation, chemical discharges, and apportionment of fish among sport and commercial interests. I have undertaken here to condense many years of personal observations and academic and applied experience into a fathomable snapshot of determinants of fish population welfare and how the scientist approaches the development of water quality criteria and fisheries management decisions.

Finally, a chapter assigned to the history of the profession and laws dealing with fisheries management. Few people have a full appreciation of the history of this field in America, which dates back to the mid-nineteenth century in a never-ending race to keep pace with growth. This chapter hopes to remedy that, along with an accounting of the myriad laws dealing with fisheries per se, their habitats, and water quality. I end the chapter, and indeed my book, with a slate of activities you may want to try as complementary avocations to the pure fun of fishing.

CHAPTER TWO

The Variety
of Fishes

Fishes were the first members of the animal kingdom to have well-developed backbones with a central nerve running through them (making them *vertebrates*). Their evolution spans 500 million years. During the first 150 million of those years fishes were on top of the ecological heap, and though many disappeared toward the end of that period (some 120 million years before the first dinosaurs), the world is still left with more than 23,000 species (Long 1995). The American Fisheries Society (AFS), our oldest professional fishing society, lists 2,412 nonextinct species in its *Common and Scientific Names of Fishes from the United States and Canada* (AFS 1991). That's a fair bit of diversity, but obviously nothing compared to that of continents with portions of their mass in the tropics.

This chapter will acquaint you with the origins of our species and some key differences among them, and serve as a primer on some elementary but indispensable terminology I found myself using habitually in chapters 3 and 4.

Origins of "True Fishes"

All fishes have a common ancestor, the "sea squirt," named after the fact that when you pick one up and squeeze it, a squirt of water comes out. The adult sea squirt is not a vertebrate, nor does it even remotely resemble an ancestor of a fish. It is a small bulbous organism, looking rather like a whitish version of a Brussels sprout. But, in its formative stage (a tadpolelike larva), it possesses a rudimentary column in its tail called a "notochord." The presence of that notochord during the embryonic stage of a fish's (or other vertebrate's) life is the reason sci-

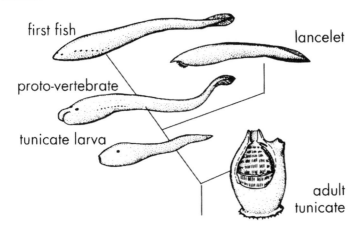

first fish

lancelet

proto-vertebrate

tunicate larva

adult tunicate

Fig. 2.1. Diagrammatic representation of how the first fishes may have evolved from protochordate ancestors. From Long, *The Rise of Fishes*, 54. © 1995 by John A. Long.

entists trace its evolution back to the little squirt. The sea squirt's notochord is resorbed during the transition from larva to adult, but the modern fish's notochord turns into a backbone. The next step up in the evolutionary chain is something called a "lancelet." This organism looks more fishlike as an adult, but the presence of a notochord during its larval stage is what really distinguishes it also as a vertebrate ancestor (Long 1995). Subsequent to the lancelet there is thought to have been a third, or missing, link, which eventually led to the development of fishes as we know them today.

The first fishes to make their appearance on earth had to get by without authentic jaws, and their notochord never did disappear or turn into a true spinal column. More than half a billion years later, though, two of their forms still endure—hagfishes and lampreys. Of three species of hagfish known to North America, one is found in the Atlantic Ocean, and of eighteen lamprey species, two (the Atlantic sea lamprey and the American brook lamprey) are found in New Jersey. None are about to take your hook. The brook lamprey is small and unobtrusive, but you may notice the parasitic sea lamprey on its way upriver to spawn. I have seen groups of them holding fast to the low dam on the north branch of the Raritan River at Route 202 by what is called their "oral disk," and swimming at my ankles in the lower Paulinskill River in Warren County during trout season. They are long and cylindrical, looking, at first glance, like eels, but if you pick one up and get a gander at its oral disk, you'll drop it like a hot potato. That disk, which I'm sure must have been the inspiration for many a horror movie, is used to suck onto the side of a large fish, rasp into it, and eat the fish essentially inside out. Having gained access to the upper Great Lakes via the Welland Canal, which let them get past Niagara Falls after 1829, they almost wiped out the lake trout and whitefish populations of Lake Michigan in the fifties and early sixties. It wasn't until the Michigan Department of Natural Resources developed a "lampricide" capable of killing the lamprey's larvae, which remain conveniently embedded in river bottoms for several years, that the destruction of lake fish stocks was thwarted.

Jawed fishes are, for the most part, much more pleasing to the eye. They come in two varieties — those whose skeleton is made of cartilage (a tough gristly substance the likes of which gives shape to your ears) and those whose frame is made of bone (*Osteichthyes*). Sharks, skates, and stingrays have the former kind of support, all other fishes the latter. The oldest sharks appeared about 438 million years ago, the oldest bony fish about 410 million years ago (Long 1995). But it wasn't until a little more than 200 million years ago that representatives of what was destined to become the most successful group of bony fishes, the *ray-finned* ones, appeared. The other two main groups of bony fishes are *lobe-fins*, which have fins comprised of rays at their edges but fleshy lobes near the body, and *lungfish*, which can breathe air out of water. Lobe-fins are found only off the coast of South Africa, and lungfish only in Africa, South America, and Australia.

This book describes cartilaginous and bony ray-finned sport fishes common to the New Jersey area. Representatives of fifteen taxonomic orders are included (taxonomy being a branch of science dealing with classification of organisms into different categories based on notochords and other characteristics). Of all of these taxonomic orders, the one that stands out as being the most diverse is that of the *Perciformes*. Perciform fishes include everything from sunfish to walleyes (freshwater); to striped bass, sea bass, blackfish, bluefish, drum, and mullet (saltwater coastal species); to tunas (oceanic species). I was going to try to figure out why, but when I read Long's (1995) book *The Rise of Fishes*, I decided not to bother. The reason? Even Long called them a "taxonomic duffel bag" of similar-looking species that have been placed in that order without anyone's really knowing if they are closely related. However, their fins have a combination of soft and spiny rays, and Lagler, Bardach, and Miller (1962) note that, unlike purely soft-rayed fish orders such as trout and herrings, their gas (or "swim") bladder lacks a connection to the esophagus. That is important (for reasons not pertinent to this book) to the taxonomist, anatomist, and physiologist. Suffice it to say that swim bladders are unique to bony fishes, enabling them to regulate their depth with minimal expenditure of energy, make sounds, or (for some of those whose air bladders are connected to the esophagus) gulp air at the surface for short periods of time when oxygen in the water is depleted.

Water—The Common Denominator

Fish rely for the most part on gills to extract oxygen from the water and give back carbon dioxide, the waste product of respiration. Gills are a series of parts packed with blood vessels exposed to the water through which these gases diffuse back and forth. Vessels in our lungs accomplish the same objective, but from air.

Besides deriving enough oxygen from the water, a fish also needs to reckon with maintenance of its internal salt balance. The blood of fish, like that of humans, has a constitution much like seawater. Unlike humans, though, fish live *in* the water, constantly soaking it up or gulping it down with their food. Saltwater species simply need to get rid of salts and water in roughly the same proportions as they take them in. They do this by expelling some salts out of specialized cells in the gills ("chloride secretory cells") and only excreting as much water as is necessary to dispose of metabolic wastes and other superfluous salts. If placed into freshwater, however, most will die because they can't cope with the amount of low salt-content water entering their bodies. If you want an animated illustration of this "osmoregulatory" enigma, try dropping blue crabs into a bucket of fresh water. You'll see a copious stream of bubbles and water coming out of

9

Table 2.1 List and Taxonomic Classification* of Fishes Discussed in This Book
(Phylum—Chordata; Subphylum—Vertebrata)

ORDER	FAMILY	GENUS	SPECIES	COMMON NAME
Class—Elasmobranchiomorphi (Cartilaginous Fishes)				
Lamniformes	Odontaspididae	*Odontaspis*	*taurus*	sand tiger
	Lamnidae	*Carcharodon*	*carcharias*	great white
	Carcharhinidae	*Mustelis*	*canis*	smooth dogfish
		Carcharhinus	*plumbeus*	sandbar shark
		Prionace	*glauca*	blue shark
Squaliformes	Squalidae	*Squalus*	*acanthias*	spiny dogfish
Rajiformes	Rajidae	*Raja*	*erinacea*	little skate
Class—Osteichthyes (Bony Fishes)				
Anguilliformes	Anguillidae	*Anguilla*	*rostrata*	American eel
Clupeiformes	Clupeidae	*Alosa*	*aestivalis*	blueback herring
		Alosa	*pseudoharengus*	alewife
		Alosa	*sapidissima*	American shad
		Brevoortia	*tyrannus*	Atl. menhaden
	Engraulidae	*Anchoa*	*mitchilli*	bay anchovy
Cypriniformes	Cyprinidae	*Carassius*	*auratus*	goldfish
		Cyprinus	*carpio*	common carp
		Luxilus	*cornutus*	common shiner
		Notemigonus	*crysoleucas*	golden shiner
		Pimephales	*promelas*	fathead minnow
		Rhinichthys	*atratulus*	blacknose dace
		Semotilus	*atromaculatus*	creek chub
		Semotilus	*corporalis*	fallfish
	Catostomidae	*Catostomus*	*commersoni*	white sucker
		Moxostoma	*macrolepidotum*	s-hd redhorse
Siluriformes	Ictaluridae	*Ameiurus*	*melas*	black bullhead
		Ameiurus	*nebulosus*	brown bullhead
		Ictalurus	*punctatus*	channel catfish
Salmoniformes	Esocidae	*Esox*	*lucius*	northern pike
		Esox	*masquinongy*	muskellunge
		Esox	*niger*	chain pickerel
	Salmonidae	*Oncorhynchus*	*mykiss*	rainbow trout
		Salmo	*trutta*	brown trout
		Salvelinus	*fontinalis*	brook trout
		Salvelinus	*namaycush*	lake trout
Gadiformes	Gadidae	*Gadus*	*morhua*	Atlantic cod
		Merluccius	*bilinearis*	silver hake
		Microgadus	*tomcod*	Atl. tomcod
		Pollachius	*virens*	pollock
		Urophysis	*chuss*	red hake

ORDER	FAMILY	GENUS	SPECIES	COMMON NAME
Batrachoidiformes	Batrachoididae	*Opsanus*	*tau*	oyster toadfish
Atheriniformes	Cyprinodontidae	*Fundulus*	*heteroclitus*	mummichog
	Atherinidae	*Menidia*	*menidia*	Atl. silverside
Scorpaeniformes	Triglidae	*Prionotus*	*carolinus*	northern sea robin
Perciformes	Percichthyidae	*Morone*	*americana*	white perch
		Morone	*saxatilis*	striped bass
	Serranidae	*Centropristis*	*striata*	black sea bass
	Centrarchidae	*Ambloplites*	*rupestris*	rock bass
		Lepomis	*gibbosus*	pumpkinseed
		Lepomis	*macrochirus*	bluegill
		Micropterus	*dolomieu*	smallmouth bass
		Micropterus	*salmoides*	largemouth bass
		Pomoxis	*nigromaculatus*	black crappie
	Percidae	*Perca*	*flavescens*	yellow perch
		Stizostedion	*vitreum*	walleye
	Pomatomidae	*Pomatomus*	*saltatrix*	bluefish
	Sparidae	*Stenotomus*	*chrysops*	scup
	Sciaenidae	*Cynoscion*	*regalis*	weakfish
		Leiostomus	*xanthurus*	spot
		Menticirrhus	*saxatilis*	northern kingfish
		Pogonias	*cromis*	black drum
	Mugilidae	*Mugil*	*cephalus*	striped mullet
	Labridae	*Tautoga*	*onitis*	tautog
	Ammodytidae	*Ammodytes*	*americanus*	Amer. sand lance
	Scombridae	*Scomber*	*scombrus*	Atl. mackerel
		Scomberomorus	*maculatus*	Span. mackerel
		Euthynnus	*alletteratus*	little tunny
		Sarda	*sarda*	Atl. bonito
		Thunnus	*alalunga*	albacore
		Thunnus	*thynnus*	bluefin tuna
	Stromateidae	*Peprilus*	*triacanthus*	butterfish
Pleuronectiformes	Bothidae	*Paralichthys*	*dentatus*	summer flounder
	Pleuronectidae	*Pleuronectes*	*americanus*	winter flounder**
Tetraodontiformes	Tetraodontidae	*Sphoeroides*	*maculatus*	northern puffer

*Based on AFS (1991)
**AFS (1991) uses genus name *Pleuronectes*. Recent evidence indicates the original genus name *Pseudopleuronectes* was correct.

them as they seek desperately to expel water invading their bodies through soft unprotected tissues in order to keep the necessary salt-to-water ratio.

The freshwater fish, having a concentration of internal salts much higher than that of its surroundings, has been designed to conserve as much salt as it can extract from its food or aqueous environment while excreting water in excess of that needed to eliminate metabolic wastes. To do this, freshwater fishes have highly developed kidney structures and bodies coated with a slimy film that limits the amount of water soaked up through their skin. The water-repelling properties of that film, plus its importance in warding off diseases and parasites, are why anglers should minimize handling of fish they intend to release. Fishes that utilize both fresh- and saltwater environments during portions of their lives are faced with the daunting task of changing their salt balance in midstream.

A General Vocabulary

Though I tried earnestly to minimize the use of technical jargon in this book, some terms having to do with a fish's body parts, habitat, and lifestyle could not be used sparingly. This section provides, in my own words, a simple glossary of the commonly used terms.

First there are terms that deal with the fish body. Since a picture is worth a thousand words, I have illustrated key components of what I will call a "holistic" fish — one that doesn't exist but which shows where everything might be if the fish possessed them in the first place. For instance, whereas all fish have "dorsal" (top), "ventral" (bottom), "anterior" (front), and "posterior" (tail) portions, in addition to gills and fins, few have "barbels" around their mouths or "adipose" fins near their tails, and sharks lack the hard "opercular flaps" that cover the gills of bony fishes. Some fish don't have separations in their dorsal fins, and the flounder's *sides* wind up being its dorsal and ventral surfaces (as chapter 4 will illustrate). You may want to dog-ear the holistic fish for future reference.

Here is some terminology worth remembering relative to a fish's lifestyle and its environment:

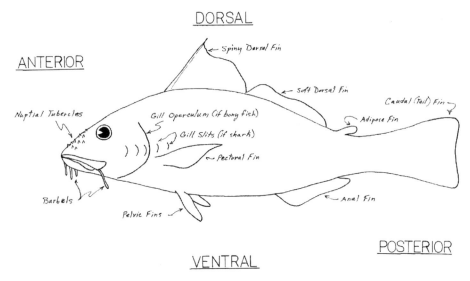

Fig. 2.2. A fictitious "holistic" fish illustrating anatomical terms of importance.

- *benthic*— of or pertaining to the bottom of the waterbody, as in benthic habitat, a benthic feeder, or feeding on the "benthos"
- *epibenthic*— of or pertaining to living on top of a substrate, such as insects on rocks, or barnacles on a piling or jetty
- *pelagic*— of or pertaining to an open-water or free-swimming existence, as in a pelagic lifestyle, a pelagic feeder, or found in the pelagic zone
- *planktonic*— of or pertaining to organisms that drift around at the mercy of the currents, such as the planktonic egg or larval stages of a fish, or a fish being a planktonic feeder (i.e., feeding on other animals or plants in the plankton community)
- *estuarine*— of or pertaining to habitats where fresh and salt waters mix (e.g., estuaries like the lower Hudson River)

That's it for now. More terms are defined later on an as-needed basis to help the reader understand either a particular species or the zonation in waterbody types described following discussion of the fishes to be found in our fresh and salt waters.

13

CHAPTER 3

Freshwater Fisheries

Altogether, fishes, their coinhabitants, and the waterbodies they live in constitute "aquatic ecosystems," defined by Armantrout (1998) as "any body of water, such as a wetland, stream, lake, reservoir, or estuary that includes all organisms and nonliving components, functioning as a natural system." If we take fish from their ecosystem, for fun, food, or profit, we have created a fishery. This chapter describes sport species and common baitfishes that are part of the area's freshwater ecosystems, and the nature of those ecosystems, along with a guideboard to many such places in New Jersey. In picking places to list on the guideboard, I focused on waterbodies with public access unless some singular experience I enjoyed on a private waterbody (such as Highland Lake, where I lived for four years) seemed helpful in making a point about a particular species.

As far as waterbodies go, I alluded earlier to the fact that from a contaminant "point source" (i.e., direct discharge) standpoint, many of New Jersey's waterbodies are far better off today than they were fifty years ago. On the other hand, continued encroachment on habitat and runoff of "non–point source" solids, nutrients, and compounds, such as oil from poorly maintained vehicles, continue to threaten these waterbodies. The New Jersey Department of Environmental Protection (NJDEP 1998a) has assessed a fair number of streams and lakes using, respectively, a biological sample index and nutrient analyses to measure ecosystem quality. They found that, of 3,815 stream miles assessed with the biological index to determine their "ability to support a healthy and diverse aquatic community," 35 percent "fully supported," 53 percent "partially supported," and 12 percent "failed

Illus. 2. The north branch of the Raritan River at Route 202 on opening day of trout season, 1998

to support" such function. Reader, beware not to misinterpret the word "healthy," however. Some habitat does not, and thus the stream possibly never could, support some of the species essential to producing a biological index high enough to be equated with full support (albeit many species forming the basis of this index, especially insects, have long been known to be intolerant of poor water quality, all other things being equal). The lake index is reflective of the fact that such waterbodies are confined and more vulnerable to contaminant build-up. Using measurements of nutrient enrichment, the state found that 33 percent of the public lakes surveyed were either threatened by, or in, a state of "eutrophication." This is defined by Armantrout (1998) as a "natural and human-influenced process of enrichment with nutrients, especially nitrogen . . . leading to an increased production of organic matter." That is, the waterbody is less clear, and may have less oxygen in summer, while not necessarily being what we tend to think of as "polluted."

Despite this slightly negative bias, let's look at the innumerable species that inhabit our waterbodies and the fishing opportunities one may enjoy.

The Fishes

According to the NJDEP Bureau of Freshwater Fisheries checklist of freshwater fishes, there are twenty-one families and eighty-four species of fish in New Jersey, including a couple of hybrids. I have chosen to describe thirty species from eight families that embrace sport fishes. Three species are "anadromous," meaning they spend their adulthood in the ocean but return to freshwater to spawn. One species is "potadromous," meaning that it swims upriver to spawn, but afterward chooses to remain in downstream fresh waters to pursue its nonconjugal life. I start with the trouts, as they have historically pumped the most money, through sales of fishing gear, licenses, and accompanying "trout stamps," into support of the NJDEP's Division of Fish, Game and Wildlife.

THE TROUTS.

To the freshwater angler who was born and bred in New Jersey, opening day of trout season is *the* day of the year—a day unlike any other, when many a youth goes sleepless with anticipation the night before. To the angler raised someplace where native, naturally reproducing populations of trout are the rule rather than the exception, well, they question our sanity. I long ago stopped losing sleep over opening day of trout season in New Jersey or anywhere else, but there is a plethora of spots right in our own backyard where one can enjoy fine trout fishing. You just need to pick your places and time your visits astutely, avoiding opening day and the first few days after restocking, in order to gain an objective angle on New Jersey trout fishing. Or, you may try one of the small headwater streams with natural trout populations described later.

New Jersey has a richer heritage of trout fishing than one might expect. The Meisselbach reel company, eventually acquired by the Airex Corporation, was based in New Jersey. The only fly reel I still own is a little brown and gray single-action Airex-Meisselbach in the

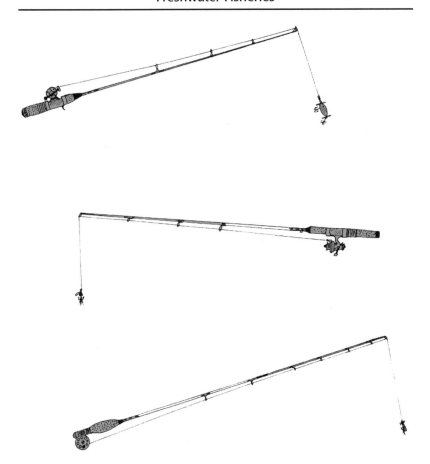

Fig. 3.1. Freshwater fishing rods and reels: (*top to bottom*) bait casting, spinning, and fly-fishing.

ultralight class. Garcia Corporation, makers of the Mitchill series of pioneering spinning reels, and the Shimano Corporation, which made my present ultralight spinning reel, were also headquartered in New Jersey at one time (see Andy Gennaro's account of New Jersey's trout fishing history in the March 27, 1997, edition of *The Fisherman*). The first complete split-bamboo fly rods produced in the United States were made in Newark, New Jersey, in the early 1860s (Trout Unlimited 1975), and we can boast of having our own fair share of the country's preeminent fishing experts, outdoor writers, and fly-tying artisans.

17

New Jersey counts four species of trout among its salmonid community. The brook trout (*Salvelinus fontinalis*, sometimes called "speckled trout") is our state fish and the only species native to the East. The brown trout (*Salmo trutta*) was introduced from Germany via a hatchery in New York State in the late 1880s (Laycock 1966). The rainbow trout (*Oncorhynchus mykiss*, formerly classified as *Salmo gairdneri*) was indigenous to the western United States but was widely introduced to the Midwest and East during the early 1900s. And finally, the lake trout (*Salvelinus namaycush*) is a species indigenous to the Great Lakes but was introduced to New Jersey's Round Valley Reservoir in 1977. The first three of these are raised in New Jersey's 50-acre, state-of-the-art Pequest hatchery in Oxford, which commenced operation in 1981. The fourth is reared in the Charles O. Hayford hatchery in Hackettstown, which pioneered the noncommercial production of trout in 1912, shifting to production of lake trout and a variety of warm- and cool-water species after the Pequest came on-line. In the fall of 1997 the Pequest hatchery dedicated 16,065 eight-inch brown trout to stocking the tidal portion of the Manasquan River in the hope that some of these fish would go to sea and return in two to three years as spawning "sea-run browns." In winter '98/99 one did just that; it returned as a 16¼-incher. While a successful sea-run population would further enrich sport fishing options, they are not expected to reproduce successfully in the silty habitat of the Manasquan. In nature, all riverine trout require clear, cool, well-oxygenated waters and clean pebbly substrates for making nests ("redds") in which to rear their eggs and larvae. When development alters those conditions, as we will see later, the trout need assistance if they are to be present in numbers sufficient to satisfy sport demand.

In waters that support successful spawning and propagation of natural populations, rainbow trout spawn in spring, brownies and brookies in fall. All three build redds, which house the fertilized eggs and resultant "yolk-sac larvae" as they develop over a one- to three-month period, after which they wiggle out into the stream and begin to feed on their own. Lake trout are different. They broadcast their eggs less carefully over rocky shoals and lake bottoms, providing no parental care.

18

Natural (vs. stocked) populations of brook trout still exist in some small streams now designated "wild trout waters," such as Flanders Brook, Van Campens Brook, and Dark Moon Brook/Bear Creek in northern New Jersey, but most trout are raised in the Pequest hatchery and stocked in what are called "trout maintenance waters." The hatchery stocks well over half a million trout every year, most of which are distributed among streams, reservoirs, lakes, and ponds from mid-March through May. A smaller portion of the total allotment is held for release into those waterbodies most popular with anglers during October, when water temperatures return to levels more tolerable to trout ($\approx$ 65°F). In addition to loss or decay of natural substrates needed to create well-oxygenated redds, which was recognized as early as 1900 and led to the decision to build a hatchery, temperatures in most New Jersey streams are too high for trout during summer. Both problems can be traced to the deleterious impact of casual land development practices in watersheds that reduce shade from tree cover and promote increased overland runoff with its complementary load of silt and other offensive matter.

In 1998 the NJDEP stocked roughly 224,000 brook, 224,000 rainbow, and 127,000 brown trout during spring, 30 percent of which were released in the two weeks preceding opening day (*Newark Star Ledger*, April 19, 1998). Fish in the initial batches average about 10½ inches in length ("standard production trout"), but one can always count on hooking a fish in the 12- to 15-inch range. In fall 1998, 45,000 standard production trout and 1,500 "breeders" in excess of 20 inches were released. Also, some 100,000 "surplus trout" smaller than the minimum size limit of 7 inches were liberated for the first time. In prior years, surplus trout were simply destroyed because they were too short for anglers to keep but too expensive to maintain in the hatchery (*Newark Star Ledger*, October 1998).

I have illustrated the three kingpins of New Jersey's traditional trout fishery. Like all trout family members, including salmon, they possess the small, fleshy adipose fin near their tails. Brook trout are bluish gray on white with wormlike markings and light spots interspersed with flecks of red. Rainbow trout resemble their name — they have a broad pink band extending from head to tail. The state record brookie (7 pounds 3 ounces) was caught in the Rockaway River in

Fig. 3.2. Trouts: (*clockwise from left*) brookie, rainbow, and brownie.

1995 (its size proving that some trout can survive, or "hold over," a fair number of summers in some of our rivers). The record rainbow (13 pounds) was caught in 1988 in Lake Hopatcong. The laker is one of New Jersey's largest trouts, the record being one taken in 1994 in Round Valley (24 pounds 14 ounces), but brown trout lunkers found in reservoirs and Lake Hopatcong are not far behind in size. Round Valley also yielded the biggest brown trout in 1995 (21 pounds 6 ounces). Lake trout are only stocked in Round Valley and Merrill Creek reservoirs. They can be distinguished from browns by their darker slate gray sides and proliferation of light yellowish spots, even on their fins, whereas brown trout sport an opposite pattern (dark spots, including some striking orange and red ones, on a lighter background). All trouts make for delicious eating, but I think the small ones just over minimum size are the best. Nothing can beat a mess of freshly caught, pan-fried little trout served next to eggs for breakfast on a chilly camping trip morning.

Every year the NJDEP issues a compendium of regulations, which should be provided when you purchase a license and "trout stamp" from your local bait-and-tackle store. This compendium includes a complete list of trout-stocked waters and the dates each is closed for restocking. It also lists wild trout waters, which are not stocked but have their own special regulations. The existence of the state compendium notwithstanding, I will mention a few locales I'm partial to.

One is the half-mile stretch of the Rockaway River on Berkshire Valley Road between Routes 23 and 15 in Oak Ridge. Being within a public water supply watershed, its banks and water quality have been preserved over the years. Opening day is not the best time to go here because the reach is too confined, and waders are a necessity to negotiate a few deep pools and/or get beyond the thick brush defending the best waters (i.e., away from the road). Other nice places include the Musconetcong River in Stevens State Park just north of Hackettstown and the stretch paralleling Route 57 south of Hackettstown (a "no-kill" reach); the "Ken Lockwood Gorge" stretch of the south branch of the Raritan upstream of High Bridge off Route 31, or the "Blewett Tract" on the Big Flatbrook downstream of Route 206 (both fly-fishing-only sections, but not "no-kill"); the Black River from Route 513 west and south of Chester through Hacklebarney State Park; the Paulinskill River upstream of Route 15 in Lafayette; the south branch of the Raritan River just west of Long Valley and, further downstream off Route 202, Neshanic Station; the north branch of the Raritan where it crosses Route 202 west of Somerville; and, given that I lived in Ocean County and the southern part of Monmouth County for eleven years, even the freshwater portions of the Manasquan, Metedeconk, and Toms rivers. Detailed accounts of trout fishing in these and many other flowing waterbodies are rendered in Perrone's (1994a) book *Discovering and Exploring New Jersey's Fishing Streams and the Delaware River*.

If lake fishing is your bailiwick, Spruce Run, Round Valley, Merrill Creek, Monksville, and Wanaque reservoirs are commendable

Fig. 3.3. An assortment of things used to catch trout in streams.

choices, as is Lake Waywayanda and some small lakes like Lake Ocquittunk in Stokes State Forest. Shore fishing works during spring
and fall, and ice fishing is popular in winter, but you will need a boat
to reach trout during summer when they go deeper to find preferred
temperatures. Sites that fail to get much publicity (since they are not
by nature trout waters), but still provide functional trouting opportunities for suburban youngsters during April, are ponds like Echo
Lake and Diamond Mill Pond, and the good old Rahway River in
Union County, which I used to reach via bicycle when Dad wasn't
around to drive me somewhere more exotic.

In the first two weeks of the official season, stocked trout are most
responsive to worms, fathead minnows, salmon eggs, spoons, and
spinners that imitate small baitfishes. Live bait is too much of a nuisance for me (I like to wade and travel light), so, depending on casting room and mood swings, I'll choose a lightweight fly outfit with a
salmon egg or an ultralight spinning outfit with something like a
Mepps, Panther Martin, Rooster Tail (three kinds of spinners), Dare-
devle, or Phoebe (two spoons). Later, when the crowds thin, the waters clear, and the trout start having to fend for themselves on natural
baits, you can start having fun with artificial flies, including nymphs,

22

other wet flies that imitate small fish or freshwater shrimp, or dry flies that mimic native bugs during hatching and mating. I've had good luck fishing the Flatbrook at Roy Bridge using a "Squirreltail" (bucktail streamer imitating a blacknose dace) fished systematically upstream in a sequence of short pulls and pauses I learned from a master of the technique on the Sawmill River in Massachusetts. I have also had success fishing an imitation of the freshwater scud (*Gammarus*), a tiny benthic crustacean important in the diet of trout in sandier glides, the likes of which may be found in the upper Rockaway, parts of the middle Paulinskill near Route 202, the midsection of the Musconetcong, the south branch of the Raritan at Long Valley and downstream of Clinton, and no doubt many other places.

Then there are the true insects and their man-made imitations. You may pick wet (fished underwater) or dry (fished atop the water) versions, depending on whether you are trying to present a fly that looks like the bug's "nymph" stage (which dominates its life, underwater), or its adult stage (which after the "hatch" and final molt flies about for a short period of time looking to mate). The three most popular taxonomic orders of insects are the stone flies (*Plecoptera*), mayflies (*Ephemeroptera*), and caddis flies (*Trichoptera*). Richey (1980) provides an elementary description of each and a state-by-state inventory of common species, the names of their imitations, when they start to hatch, and the duration of the species' hatching period in his book *The Fly Hatches*.

Stone flies spend the first year or two creeping about as nymphs, eventually crawling out of the water onto rocks and branches to emerge as adults, mating and depositing their eggs on the surface of the water. Stone flies have two tails and range in size from ⅛ to 2 inches. In early April, if you choose to try a dry fly, your best bet may be the "Early Brown Stonefly." Toward the middle of April and (for some species) through summer, caddis and mayflies begin hatching.

There are four types of mayfly nymphs—crawlers, clingers, swimmers, and burrowers. When they are ready to start a new generation they swim to the surface, shed what is called their "nymphal shuck" (i.e., hatch) to become "duns," which fly to shoreline branches, and then wait up to twenty-four hours to perform their final molt. Most mayflies have three tails, but there are some common exceptions.

Having completed the final molt, they are now called "spinners." Their wings are generally more clear, and they can be seen flying about seeking to mate, only to die right after their mating ritual. Imitations of both airborne forms exist. Common patterns in New Jersey are the "Brown Spinner," the "Red Quill" (a male dun), the "Quill Gordon" (and its swim-up emerger the "Hare's Ear" wet fly), the "Light Hendrickson" (a female dun), the "Light Cahill," and the large "American March Brown."

Common caddis fly patterns include the "Little Black Caddis," the "Dark Caddis," and the "Green Caddis." Caddis flies mate overland and may live long enough (ten days) to mate two or three times before dying, but the most interesting thing about them is what they do as nymphs. Depending on the species, some construct little tubelike cases of gravel or tiny bits of wood cemented together by a secretion from the nymph, others erect a netlike funnel facing upstream to catch food. Pick up any rock in a clear, swift riffle (these organisms don't do well in slow, muddy waterbodies), and you should be able to find a caddis case or two, maybe along with some mayfly and stonefly nymphs that didn't jump off as you retrieved it.

For additional reading, besides Richey's (1980) book, try Ernest Schwiebert, Jr.'s (1955) *Matching the Hatch*, the New Jersey Council of Trout Unlimited's (1975) *New Jersey Trout Guide*, Caucci and Nastasi's (1984) *Instant Mayfly Identification Guide*, or Bergman's (1938) classic *Trout*.

BASSES, SUNFISHES, AND CRAPPIES.

These species, members of the family Centrarchidae, are the most obvious "nest builders" of the sport fish world. They all spawn in spring and early summer, with the male in each case building and guarding the nest. They are all fun to catch and (with the possible exception of one species) good to eat. Given your druthers, you can catch them with spin- or fly-fishing ensembles, adding bait-casting equipage to that list for largemouth bass and a cane pole for sunnies. Swimming and surface plugs, spinners and spoons, wet and dry flies, bucktails, cork or soft plastic "poppers" that imitate small frogs and mice, colorful plastic worms, and bobbers with night crawlers or baitfish all have

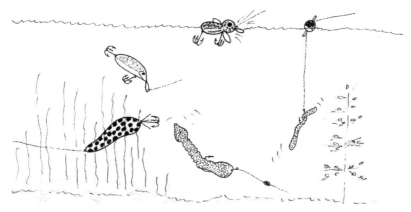

Fig. 3.4. A potpourri of largemouth bass baits.

their niche in the centrarchid angler's bag of tricks. Centrarchids can be found throughout the state in ponds, lakes, reservoirs, and mid- to downstream reaches of rivers, although the smallmouth bass is more picky about habitat, and the rock bass just doesn't seem to be as common in New Jersey (in my experience) despite its reputed tolerance of assorted environmental conditions.

The largemouth bass (*Micropterus salmoides*) is native to the South, where it was first noted in Florida in 1562, but has been "naturalized" almost everywhere, here beginning with plantings in 1871 (Mac-Crimmon and Robbins 1975). Largemouth bass inhabit lakes and reservoirs as small as Surprise and Echo lakes in Union County, to those as large as Lake Hopatcong in Morris County and the Manasquan Reservoir in Monmouth County, in addition to riverine habitats like the Delaware River and the Delaware and Raritan (D&R) Canal.

They commonly feed near shore among fields of lily pads, and when it's time to think about spawning the males make 2- to 3-foot diameter nests in fringing areas, which they defend resolutely. So resolutely, in fact, that after being caught and released following a territorial anxiety attack they make a beeline back to their nests, only to make the same mistake again! This behavior, which I observed regularly in May or June fishing an orange and black spotted Flatfish (or, for that matter, just about anything) in Union County's Surprise Lake

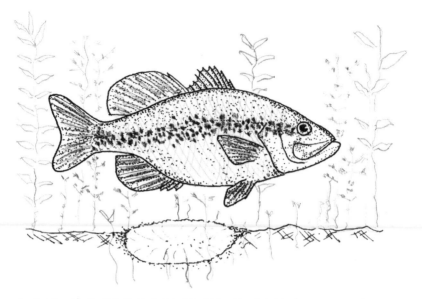

Fig. 3.5. Largemouth bass over an early stage of a nest in progress.

and Nomahegan Park Lake decades ago, is not necessarily conducive to perpetuation of their populations. It is also chief among reasons for the NJDEP's requiring us to release them immediately between April 15 and June 15.

Even so, studies of bass survival during competitive tournaments across North America have shown that the act of hooking, playing, and landing the fish results in an estimated 6.5 percent initial mortality rate (e.g., if they take the hooks too deeply and/or are just dumped back in the water after a long fight) and a delayed mortality rate averaging 23.3 percent (Wilde 1998). The average angler can try to minimize delayed mortality by wetting his or her hands first to reduce the amount of protective slime scuffed off the fish, grasping its lower jaw ("lipping" it), and holding it vertically (i.e., letting it dangle) rather than horizontally to remove the hook.

Should the male largemouth bass survive this noble time of making his nest and guarding his thousands of eggs and larval offspring, most of which fail to reach their juvenile stage despite his efforts, he may become much more wary and difficult to catch with age. Back in

1962 when Route 23 north of Newfoundland was still a cow path and nocturnal sweethearts would spin tales of the maniac with a hook for a hand, I spotted a huge (at least 2-foot) largemouth bass sunning itself at the surface of one of the pristine little lakes between Stockholm and Franklin, which have long since been privatized and developed. It was a midsummer day and I had the perfect vantage and casting point atop a large boulder some 50 feet away from this prize. I remain convinced that I was beholding a record largemouth, though, if you're specifically coveting a state record, biologists suggest fishing more southern waters draining toward the Delaware River. Two lakes in particular, Assunpink (Monmouth County) and Parvin (Salem County), have been designated "Trophy (or Lunker) Bass Lakes" by the NJDEP. At these, there is a size limit of 15 inches and a bag limit of three fishes.

I baited my hook with a juicy night crawler that I had prevented from procreating on the front lawn of my house in Union the night before, set a 3-foot dropline from my red-and-white bobber, and made a perfect cast and soft touchdown about 20 feet beyond the fish. He didn't move, and I slowly drew my bait to within 5 feet of his senses, leaving it there to seduce him. Sure enough, in five minutes he ducked underwater, and my bobber began to move. I had heard that a largemouth doesn't always take the hook when it starts mouthing the bait. So with heart pounding, I waited for what seemed to be an eternity but was probably just five minutes, to set the hook once my bobber finally took a sustained and uniform 4-foot dive. Unbelievably, without any resistance, my bobber simply popped to the surface with the worm intact, leaving the record for someone else to claim in 1980 (10 pounds 14 ounces, Menantico Sand Wash Pond in South Jersey). That size bass has to be pushing fifteen-plus years.

Fortunately, adult bass in the 12- to 14-inch size range are less difficult to catch during summer. Though worms are always reliable, I personally prefer the action created when a bass comes from under a soft, weedless plastic spinning popper or a colorful feathery cork fly popper fished among the weeds. "Bass buggin'"—very exciting.

Although most small ponds throughout the state provide good bass habitat and fun fishing, here are some public places where you should be able to find robust largemouths — some based on my own experience, others listed by Perrone (1994a, 1994b) and the NJDEP (1994, 1998c) (* = stocked as recently as 1997; + = boat ramp):

Delaware River — Hunterdon through Salem counties +++++
Atlantic County — Lenape Lake*+
Bergen County — Scarlet Oak Pond
Cumberland County — Maurice River, Menantico Creek, and Menantico Sand Pond+
Essex County — Weequahic Park Lake*+
Gloucester County — Mullica Hill Lake
Hunterdon County — Spruce Run Reservoir*+ and Amwell Lake
Mercer County — D&R Canal
Monmouth County — Assunpink Lake+ and Manasquan Reservoir+
Morris County — Kays Pond, Lake Hopatcong+, and Lake Musconetcong+
Ocean County — Mill Pond*
Passaic County — Monksville Reservoir+
Salem County — Harrisonville Impoundment #2 and Parvin Lake
Warren County — Furnace Lake+ and Merrill Creek Reservoir+

For up-to-date news and dialog, try checking the wmi@wmi.org/bass fishing home page.

The smallmouth bass, *Micropterus dolomieu,* was described by Henshall (1881, cited by Eddy and Underhill 1974) as being "inch-for-inch and pound-for-pound, the greatest game fish that swims." I don't know about that, but certainly the 17-incher I caught in April 1998 on the south branch of the Raritan at Neshanic Station was no slouch. I caught him or her (I didn't squeeze the fish) by chance.

My day started at 0600 hours from Weehawken (my current residence) en route to the north branch of the Raritan, where I intended to spend the morning fishing for trout. Unfortunately, the day I picked was closed for restocking. After a brief period of dejected contemplation of nearby options, I hit the road again and spontaneously took a left turn toward South Branch, New Jersey, which I intuited must

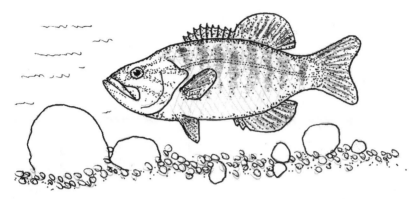

Fig. 3.6. Smallmouth bass.

be a place where the south branch of the Raritan had to pass. Following my nose, I wound up at Neshanic Station.

A dozen or so anglers were positioned below a low-head dam proximal to the parking area, flinging worms or salmon eggs at whatever trout might be dull-witted enough to bite. Not being one who enjoys elbow-to-elbow collectives, I chose the high road upstream of the barrier, which seemed to offer agreeable glides with no human beings. A half mile upstream I started tossing a yellow Rooster Tail spinner, one which normally produces dependable catches of trout, cross- and upstream of me. No hits after a half hour. So I switched to a little Mepps spinner with a pastel blue and pink blade and on my first cast got results. In two successive casts I hooked, landed, and released two beautiful 12-inch brook trout. Then, having promised my wife, Mia, I'd be home about 1 p.m., I started wading downstream toward the parking area.

My last cast was against the wall of the old mill (which is now a condo or something) above the low-head dam. As my Mepps swung downstream into the 4-foot pool above the dam, it stopped abruptly, much as it would have had it snagged an underwater log. Wading deeper to retrieve this little three-dollar investment, I jiggled the tip of my rod and the thing started moving. A half hour later, after a grand battle of wits matching its will to survive and my competency to manipulate a drag and keep it from either snapping my 2# test line or going *over* the dam, I landed the most handsome smallmouth of my life. After carefully beaching it and removing the hook from its

Illus. 3. Neshanic Station on the south branch of the Raritan River. From Cawley, James, and Cawley, Margaret, *Exploring the Little Rivers of New Jersey*. Copyright © 1971 by Rutgers, The State University. Reprinted by permission of Rutgers University Press.

jaw, I spent the next fifteen minutes holding it face-forward into the stream to revive it by getting the oxygen flowing through its gills again, and on its way to fight (or spawn) again another day, thus lessening the odds of delayed mortality I spoke of earlier.

My smallmouth probably would have weighed in at about 4 pounds, the state record being a 7-pound-2-ounce brute caught in 1990 in Round Valley Reservoir. Coble (1975) calculated, using growth rate data reported for a variety of populations, that smallmouth bass average 3.7 inches in length after one year, 9.2 inches after three years, 12.7 inches after five years, and 16.9 inches after nine years. With an average weight of 1 pound at a length of 13 inches (Lagler 1956), even though growth rates slacken with each advancing year, you get a feeling for how old a 7-pounder might have been.

Smallmouth bass are native to the Great Lakes and Mississippi drainage. They entered the Great Lakes during the late Pleistocene glaciation period 20,000 years ago, were first noted in the Saint Lawrence River in 1664, and were naturalized in our area circa 1870 (MacCrimmon and Robbins 1975). Though both wear a combination of greenish and creamy colors, the smallmouth may be distinguished from the largemouth by its series of vertical green bars versus a single head-to-tail longitudinal one, and the fact that the posterior end of the smallmouth's upper jaw doesn't extend beyond the back of its eye.

Unlike largemouths, smallmouths tend to be found either in flowing waters with sandy bottoms or similar habitats in lentic water-bodies such as our reservoirs. Most of the time you can catch them on spinners or minnows, but, in the words of Eddy and Underhill (1974), it is "impossible to give explicit directions for catching smallmouths because only the individual bass knows what it wants and what it will do." Some places you may expect to find them include the Delaware River from Hunterdon County to the Water Gap; Monksville, Merrill Creek, Round Valley, Boonton, and Wanaque reservoirs; Swede's Lake in Burlington County; and the mid to lower reaches of the Rockaway, north and south branches of the Raritan, Paulinskill, and Big Flatbrook rivers.

The rock bass, *Ambloplites rupestris*, is a meddlesome little creature that you might—but wouldn't go out of your way to—catch. It is dark brown and looks kind of like it belongs in this family, but it remains fairly small all its life and can't crow of being a great "panfish" (a reference to how you cook them). Still, you should know about it,

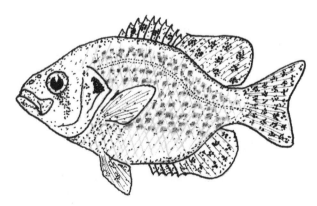

Fig. 3.7. Rock bass. In real life its eyes are red.

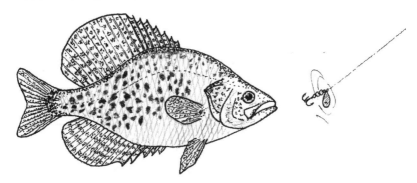

Fig. 3.8. Black crappie about to take the bait.

especially if you catch something about 8 inches long and it has big red eyes. That's the rock bass. The state record is a 1-pound-5-ounce behemoth taken in 1982 in the Saddle River in Bergen County. It is relatively nondiscriminatory (as basses go) in its selection of nesting sites, often building nests in areas laden with weeds, roots, and other debris (Lagler 1956). It feeds on small crayfish, insect larvae, small fish, and snails, the latter of which frequently cause it to be infected with "black spot," a stage in the life of parasites that use the snail and rock bass as intermediate hosts en route to their ultimate hosts, wading birds. Black spot appears as unseemly looking infestations in the fish's skin, fins, and gills. In fall these blackheads may in fact be molluscan larvae called "glochidia," which I first saw on rock bass in the Red Cedar River outside of East Lansing, Michigan. Anyway, I just catch them and release them on ultralight spinners usually aimed at smallmouth in places like Round Valley Reservoir and the downstream reaches of the Big Flatbrook, Paulinskill, and Black rivers.

Black crappie, *Pomoxis nigromaculatus*, is a wonderful sport fish I would go out of my way to catch. They readily attack a spinner, they are a great panfish, and they get pretty big (the state record being a 4-pound-8-ounce fish taken from Pompton Lake in 1996). They are easily distinguished from their relatives by their mottled coloration and sail-like posterior dorsal fin. I have caught many large ones in Mom's (private) backyard lake in Bricktown's Holiday City retirement community, well after I caught those pictured on my "stringer" circa 1955 in Surprise Lake. But the best crappie fishing I've seen anywhere in New Jersey is in Lake Waywayanda, a waterbody near the

New York State border north of Highland Lake in Sussex County that you must access via an offshoot of Route 94/511 just south of Warwick, New York. Other waterbodies that are reputed to offer good crappie fishing are Deer Path Park Pond in Hunterdon County, Farrington Lake in Middlesex County, Assunpink Lake and Manasquan Reservoir in Monmouth County, Riverview Beach Pond in Salem County, and Allamuchy Pond in Warren County.

Last but not least, the omnipresent sunfishes—two in particular. Both were described by Lippson and Lippson (1997) as bright, saucer-shaped, sparkling multicolored fishes, which is true. The bluegill (*Lepomis macrochirus*) achieves a larger size, all things being equal, than the pumpkinseed (*Lepomis gibbosus*), but both make terrific panfish. The bluegill has a black "earflap" at the back of its gill cover and a black blotch near the base of its dorsal fin. The pumpkinseed has an orange crescent around its earflap and blue and yellow stripes on the lower portion of its head. They inhabit all sorts of waterbodies, including streams and rivers where the water velocity is not too fast and

Illus. 4. The author with a stringer of black crappies taken from Surprise Lake, ca. 1955

Fig. 3.9. Bluegill (*left*) and pumpkinseed sunny whose eyes may be bigger than their stomachs.

the bottom substrate not too rocky. Their nests, usually containing an adult or two preceding or during spawning, are easily spotted within yards of the shoreline during May and June. Some of these nests may be occupied by bluegills, others within reach containing pumpkinseed, and they are known to interbreed and create hybrids.

By late August, if the water is clear enough, you may even be able to see congregations of inch-long young-of-the-year milling about. They can grow to about 1¼ inches by the end of their first year, usually maturing in their second year of growth, and may reach 9 inches (still less than a pound) by the time they are ten years of age (Lagler 1956), but, because of their high reproductive capacity, their growth rates are as variable as the habitats they occupy. In ponds or lakes without enough predators like largemouth bass or pickerels to crop them naturally, they become stunted (relatively small for their age) due to competition for food. In well-managed ponds, and/or ponds that haven't been developed and fished too much, they may be larger, age-for-age, than average. Farm ponds are perfect examples. The state record bluegill (3 pounds) was caught in 1990 in a farm pond in Pennington (Mercer County), and the record pumpkinseed (1 pound 8 ounces) was taken in 1987 from a farm pond in Burlington County. So . . . if you can, make friends with a farmer or local home owner with a private pond!

Both species have quite opportunistic feeding habits. They will eat amphipods, terrestrial and both nymph and adult stages of aquatic insects, and small fish. I have caught them on everything from as ridiculous an artifact as a half-inch rubber spider filled with an oily stuff serving as its own "chum" (haven't seen those in decades) to a small wet fly, to a worm or a spinner. Once again, my favorite form of fishing for "sunnies" is with a fly rod and tiny cork popper with a short tail and hackle (muffler antecedent to the tail formed by wrapping a feather around the shank of the hook).

As noted earlier, sunnies are everywhere the water quality and spawning substrate permit them to get by, making them a perfect choice for teaching kids to fish no matter where you live. They are the cardinal candidate in urban ponds and a neat species to go after in rural areas. Quinn (1997) mentions the fact that bluegills and pumpkinseed are both found once again in the upper reaches of the Meadowlands in Losen State and Overpeck Creeks, and here are a few more possibilities I have chosen from my own experience and information compiled by the NJDEP (1994) and Perrone (1994b):

Atlantic County—Heritage Park Pond
Bergen County—Ramapo Lake and Wooddale Park Lake
Burlington County—Mirror Lake
Camden County—Newton Lake
Cumberland County—Union Lake
Essex County—Branch Brook Lake and Weequahic Park Lake
Gloucester County—Swedesboro Lake
Hunterdon County—Amwell Lake
Monmouth County—Manasquan Reservoir
Morris County—Budd Lake and Lake Musconetcong
Ocean County—Colliers Mill Pond
Sussex County—Lake Hopatcong and Lake Waywayanda
Union County—Surprise Lake

Don't restrict yourself to these places though, because sunnies are all over.

THE PERCH FAMILY.

The AFS (1991) lists 149 North American species in this family, although there are only two you can expect to catch with rod and reel in our area. They are the yellow perch (*Perca flavescens*), which is fairly common, and the walleye (*Stizostedion vitreum*), which is not native to New Jersey but is stocked in Lake Hopatcong, Greenwood and Swartswood lakes, and Monksville and Canistear reservoirs (all in northern New Jersey). One other species is of sport size and value, the sauger, but it is endemic to the Great Lakes and, unlike the walleye, has not been stocked here. Of the remainder, 140 species are classified as "darters," small bait-sized specimens that are fascinating and in some cases beautiful fishes, but which are neither caught by anglers nor sold as bait. Of those 140 darters, only three are listed among fishes of New Jersey by the NJDEP. They are the swamp, the tessellated, and the shield darters. Though not among New Jersey's

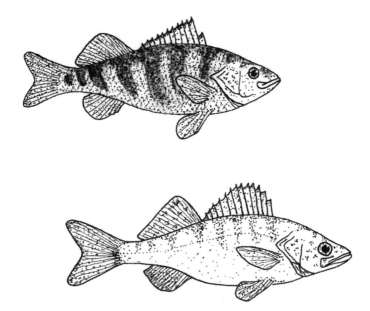

Fig. 3.10. Yellow perch (*top*) and walleye.

Percidae, five species of "logperch" are also counted among this family in North America. They too are small and interesting critters, but describing them goes beyond the scope of this book.

The yellow perch is yellow with a series of greenish gray vertical bands along its sides. It is most common in lakes and some ponds, but may occasionally be caught in streams. Like sunnies, yellow perch grow larger in some waterbodies than others. I have caught beautiful big ones (say between 10 and 14 inches in length) in Highland Lake and Lake Waywayanda in Sussex County and in Mom's former private lake in Holiday City (Ocean County), while in addition catching smaller examples in numerous other places in New Jersey. Alternative waterbodies reputed to offer good fishing for yellow perch are Lake Hopatcong (Morris and Sussex counties), Tuckahoe Impoundment #3 in Cape May County, Ramapo Lake in Bergen County, Weequahic Park Lake in Essex County, and Pompton Lake in Passaic County. The current state record yellow perch (2 pounds 6 ounces, possibly over ten years old) was caught in Holiday Lake (that may be the one in Holiday City, which is private), but the state lists a 4-pound-4-ounce yellow perch as the "historical record" (taken in 1865 from Crosswick's Creek, which enters the Delaware just below Trenton). Yellow perch can be caught on spinners and worms from spring through fall or by fishing shiners or herrings through the ice during winter. In fact, though I now prefer skiing in winter, the yellow perch is one of the most popular species among ice fishermen (a small but dedicated and tight-knit group). Yellow perch are also very adaptable to fairly salty waters, being called the "best-known freshwater fish of Chesapeake Bay" by Lippson and Lippson (1997). They note that the yellow perch has become so acclimated to these waters that in Chesapeake Bay they behave, from a spawning standpoint, more like the white perch described in my next chapter.

The yellow perch does need freshwater habitat and lots of submerged, rooted aquatic vegetation for successful spawning, however. They are spring spawners, but instead of building nests, they rely on a tactic that envelops their eggs in a gelatinous, accordion-patterned, and ropelike mass, through which the individual eggs are visible. No parental protection is provided (Dix 1995). Take a look into the water in some weeded areas of a pond near you in late April, and you might see some of these tubes, which can measure a foot or two in (stretched-out) length.

Yellow perch make excellent table fare, but sometimes they don't look too palatable due to black-spot infestations caused, like those infesting rock bass, by the larvae of parasitic flatworms that use the perch as a go-between from snails (which perch also like to eat) to their final hosts, fish-eating birds.

I have never caught a walleye, nor for that matter fished for them by rod and reel. Once I've completed this book I hope to go walleye fishing in either the Delaware River or one of our notable lakes or reservoirs, but till now my only direct encounters with walleye were in Massachusetts, where I caught them in gill nets in an "oxbow" lake off the Connecticut River, and in northern Minnesota's St. Louis River, where I captured one in a hoop net. Walleye have only recently (ca. the seventies) become popular in New Jersey, but, according to Robert Papson in the January 1998 issue of *New Jersey Fish and Wildlife Digest*, its overall popularity makes it the fastest growing sport fishery in the country. In places like Minnesota, where they are indigenous, opening day of walleye season has been likened to opening day of deer season (Eddy and Underhill 1974).

Walleye have a bigger than usual eye (hence their name); they don't show the very apparent vertical bars of the yellow perch; and they have large black splotches at the base of the last three spines of their (anterior) spiny dorsal fins. Walleye were stocked in Greenwood Lake and Lake Hopatcong during the early 1900s, but it wasn't until Monksville Reservoir was completed in the late eighties that the state really started mounting an effort to establish the fish in New Jersey. As noted in Mr. Papson's article in the *Digest*, since 1990 they have been stocked regularly in Lake Hopatcong, Greenwood Lake, Swartswood Lake, and Monksville and Wanaque reservoirs—all in the northern part of New Jersey (Sussex and Passaic counties). The upper Delaware River apparently remains the best and most well-established walleye fishery. In January 1997, *The Fisherman* noted that they were regularly being caught on "fallfish" there (see minnow section). Walleye are stocked as fingerlings, averaging 1.8 inches, and as advanced fingerlings, averaging 3.5 inches in length. The state record (13 pounds 9 ounces) hailed from the Delaware River in 1993. Eddy and Underhill (1974) point out that the walleye's white flaky meat makes for great eating—either fried, broiled, stuffed, or baked—and that in the Midwest they are typically served under the alias "walleyed pike." I've tasted walleye there and can say it's true what they say.

PICKEREL, PIKE, AND MUSKIES.

The "duckbill" family (Esocidae) has a unique look with those heads and fin arrangements. The pectoral fins are underneath the gill covers; the pelvics are beneath their midsection; and they have that single, lone soft dorsal fin, positioned all the way back toward the tail. Although their coloration and body markings differ, the surest way to tell them apart is by examining their heads.

Chain pickerel (*Esox niger*), the most common member of the family in New Jersey, but the smallest one illustrated here, have fully scaled gill covers (opercles). They also have a distinctive chainlike pattern of darker greenish markings on their somewhat lighter greenish yellow bodies. Neither muskellunge (*Esox masquinongy*) nor northern pike (*Esox lucius*) have scales on the lower halves of their opercles, but

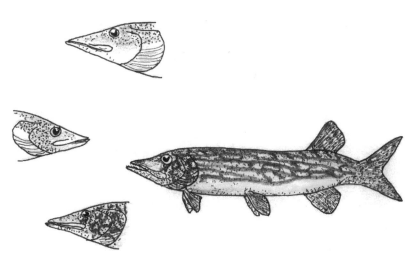

Fig. 3.11. A whole chain pickerel and (*from bottom to top*) heads of the pickerel, northern pike, and muskellunge, showing differences in cheek and gill cover scaling.

whereas the pike's "cheeks" are always fully scaled too, the lower half of the muskie's cheeks are usually barren of scales. Notice I said "usually." The definitive test of muskie versus northern pike is the number of small pores under each of their lower jawbones ("mandibular" pores). Northern pike never have more than five pores per jawbone, whereas muskellunge always have more. Since the two hybridize readily, and in fact are intentionally crossed and stocked out of the Charles O. Hayford hatchery, these "tiger muskies" sometimes feature contradictory characteristics. All have muskie body patterns, ordinarily dark spots or oblique bars on a lighter brownish green or even silvery background, but the majority retain the cheek scaling and number of mandibular pores of the northern.

The northern pike is the second largest of the three. They spawn in shallow, flooded marshy areas contiguous with streams and lakes during spring, females releasing tens to hundreds of thousands of eggs (depending on the size of the female), which are fertilized by males, sink, stick to the bottom ("adhesive" eggs), and are summarily left by the parents to fend for themselves. After hatching, the pike can grow at a rate of one-tenth of an inch per day, reaching 7 to 12 inches (½ to ¾ pound) by the end of its first summer (Lagler 1956). Based on a quick eyeballing of data presented by Carlander (1969), which cover many habitats and latitudes, the average northern pike may be about 28 inches (4 pounds) by the end of its fifth summer, and 37 inches (10 pounds) in its tenth year. Pike stocked by the NJDEP average 4.7 inches in length. The New Jersey state record "northern" (30 pounds 2 ounces, possibly about twenty years old) was caught in Spruce Run Reservoir in 1977, but, according to Lagler (1956), there are old European reports of pike weighing well over 100 pounds, and zoo (or aquarium) chronicles telling of pike that had been maintained for seventy-five years!

The northern pike's sides and back are bluish green with a profligate collection of small light spots, although juveniles have a series of light bars that later break up into spots, and its belly is white or creamish. The pike's mouth is full of very sharp teeth, which, when worn out or broken, are replaced by others waiting their turn alongside the larger rows (Eddy and Underhill 1974). You don't want to "lip" any one of these species, which is something even an experienced angler must remember. I recklessly pulled a small pickerel out of Mom's lake that way just a few years ago, ending up with a bloodied thumb.

Due to their voraciousness, northern pike have been alternately vilified by trout fishermen, and indeed consciously removed by trout enthusiasts over the years, or lauded by warm-water enthusiasts as being *helpful* to centrarchid populations by thinning their ranks and preventing overcrowding and stunting. Quoting from Walton (1936, originally published in 1653), "The mighty Luce or Pike is taken to be the Tyrant, as the Salmon is the King, of the fresh waters." More colloquially, the "Luce" (after its species name) may be called the freshwater barracuda. My first northern took a Mepps in the lip just after I flicked the spinner off a rock while wading and casting upstream toward a pool in Fish Creek, Michigan, in 1966 where I thought a lunker brown trout might be harbored. That pike must have been eating its weight in trout every day, but it was an unexpected thrill for me to fight and land a species I had only seen previously in jars of formaldehyde.

Northern pike are found naturally in cool to moderately warm and generally weedy rivers, lakes, and ponds (Lagler 1956). Northerns are taken regularly, throughout the year, in our big lakes and reservoirs using live shiners. *The Fisherman* reported pike being taken with shiners during March and April 1997 in Spruce Run and Round Valley reservoirs, and also with shiners through the ice in Budd Lake, Lake Hopatcong, and Spruce Run Reservoir in January 1999. The state has stocked northern pike in these waterbodies, as well as Deal Lake (Monmouth County); Farrington Lake (Middlesex County); Pompton Lake and River (Passaic County); Cannistear Reservoir, which you need a permit from the water purveyor to utilize (Sussex County); the Maurice River (Cumberland County); and the Millstone River (Middlesex and Somerset counties). The March 27, 1997, edition of *The Fisherman* carried a secondhand report of a 2-foot muskie being caught near Zarephath in the Millstone River. Based on stocking history, my guess is that the fish was most likely either a pike or a tiger muskie.

Northern pike are bony, having an extra set of Y-shaped bones above their regular ribs, but, in Eddy and Underhill's (1974) words, have an excellent flavor and are delicious "stuffed, seasoned and baked."

Muskellunge have been stocked in the Delaware River and Mountain Lake (Warren County), Greenwood Lake (Passaic County), Lake Hopatcong (Morris and Sussex counties), and Spruce Run Reservoir (Hunterdon County). They average 9.3 inches at stocking time, whereas two size classes of tiger muskies are stocked (fingerlings, 3 inches long, and advanced fingerlings, 8.6 inches long). Tiger muskies have been stocked in the latter two waterbodies, as well as Union Lake (Cumberland County), Shenandoah Lake (Ocean County), Carnegie and Mercer lakes (Mercer County), Shadow Lake (Monmouth County), the D&R Canal (Middlesex and Somerset counties), the lower Delaware River, and Furnace Lake (Warren County).

Muskellunge can attain enormous proportions and old ages, reportedly reaching 8 feet in length and well over 100 pounds in old Europe (Lagler 1956). Our state record, probably well over twenty-five years of age based on Ruppel's (1994) piece on Delaware River muskies, is a 42-pound-13-ounce trophy taken through the ice from Monksville Reservoir in 1997. They grow fast, attaining 7 to 9 inches after one season's growth, 13 to 16 inches after two, 26 to 30 inches after five, and 3 to 4 feet after nine (Lagler 1956). They are reputed to eat anything from ducklings to muskrats, but usually confine their diet to other fish. Muskies can be caught near shore using live shiners or herrings as big as 8 inches, by trolling or casting active spoons along weed beds, or by casting spinners or plugs near natural or man-made structures from the banks of the Delaware (Ruppel 1994). Here is a subset of Rowe's (1993) tips for catching muskies:

- Muskies can be caught from boat or shore as they lurk in deeper pools hidden in the bushes or rocks to ambush their prey.
- They are very wary and often will strike only after several passes.
- Since they are loners, it's unusual to get more than 1 in an area.

- Usually caught by fishermen fishing for other fish, so their size and sharp teeth are too much for the fisherman who isn't prepared [e.g., with a steel leader].

The same should be said of the tiger muskie, which is reported to suffer lower natural mortality and display greater vitality ("hybrid vigor") than either parent. It too may be caught throughout much of the year, including under the winter ice. A 37-incher (eight to nine years old, based on data in Carlander 1969) was reportedly taken that way from Greenwood Lake in January 1999 (*The Fisherman*, January 28, 1999). The state record tiger (29 pounds, conceivably at least fifteen years of age) was caught in 1990 in the Delaware River from a bank of the Washington Crossing State Park (Ruppel 1994). In general, "true-strain" muskellunge are found in the upper Delaware from Frenchtown to the New York State border, while tiger muskies are found further downstream, to and including some of the major tributaries in Mercer, Burlington, Camden, and Gloucester counties (Ruppel 1994).

The chain pickerel is the most universal of the Esocidae in New Jersey, being found from the lakes and portions of rivers in the north, straight through the state's waist between Trenton and Perth Amboy, and throughout our southern counties and the cedar waters of the Pine Barrens. Chain pickerel are not stocked by the state. They simply move into shallow weedy areas sheltered from currents during spring, scatter their eggs (which are adhesive) over soft substrates, and desert them. The surviving young feed voraciously and grow very rapidly, reaching (according to a Lake Hopatcong study cited by Carlander 1969) 15 inches at age two, 21½ inches at age five, and 26 inches at age seven. The current record for a chain pickerel in New Jersey, 9 pounds 3 ounces, was docketed in 1957, suggesting that pickerel growth and longevity have for some time been ecologically much more limited than they were then. Still, there are ample opportunities to catch chain pickerel in the 2-foot class.

The chain pickerel is every bit as hawkish as its larger relatives. In Highland Lake once, a huge pickerel came after and engulfed a bluegill that was already on the hook of my fly popper, an act also ascribed to muskies (Rowe 1993). It released the sunny as I lifted it out of the lake, waiting perhaps for a more rewarding test of esprit at such time as I was actually seeking northern pike or pickerel trolling a Rapala plug slowly along the edge of the weeds from the stern of my canoe. I had a similar experience in summer 1998 in the Rockaway River at Oak Ridge. A sunny that had taken a spinner I was using for hold-over trout was chased madly about my legs by a big pickerel till I rescued the little thing by lifting it out of the water. On a previous cast, my suspicions that a pickerel must be about were aroused when, in one hard slashing hit, I lost that little pink and blue spinner I had caught my smallmouth bass with in April. Like other members of the family, chain pickerel attack live shiners as readily as they do spoons and plugs, and can be caught year-round.

Examples of good pickerel waters are the Mullica River (Atlantic County), Ramapo Lake (Bergen County), the Wading River (Burlington County), the Tuckahoe River (Cape May and Atlantic counties), Bennets Mill Pond and the Maurice River (Cumberland County), Amwell Lake (Hunterdon County), the South River (Atlantic County), Assunpink Lake (Monmouth County), Forge Pond (Ocean County), Lakes Musconetcong and Hopatcong (Morris County), Pompton Lake (Passaic County), and Lake Waywayanda (Sussex County).

BULLHEAD CATFISHES.

Although this family (Ictaluridae) is officially recognized by the AFS as "bullhead" catfishes, it includes a large and popular sport fish most anglers would not think of when the word bullhead comes up in fishing conversation, i.e., the channel catfish (*Ictalurus punctatus*). Fish that most of us think of as bullheads are smaller, less streamlined, brown or black fishes that occupy shallow ponds and average maybe 10 to 12 inches. Their genus is *Ameiurus*, and I consider two species to be the most common, brown (*A. nebulosus*) and black (*A. melas*) bullhead.

All three share some common characteristics, however: no scales, adipose fins (like trout), stout spines in their dorsal and pectoral fins, and barbels around their mouths, two of which (those at the hind ends of their upper jawbones) make them look like they have whiskers. "The Barbel is so called, says Gosner, by reason of his barb or

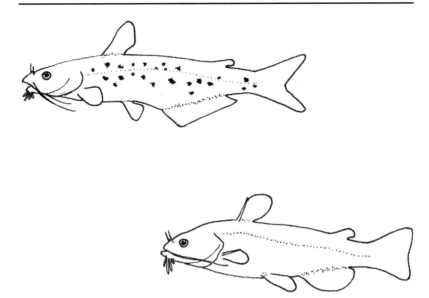

Fig. 3.12. Channel catfish (*top*) and bullhead.

wattles at his mouth, which are under his nose or chaps" (Walton [1653] 1936). All catfish are also what Eddy and Underhill (1974) call "tenacious of life" (i.e., they can stand, to a point, environmental conditions many other sport species cannot). As adults, catfish are omnivorous nocturnal feeders, and they in turn make for excellent nocturnal human dining. I've caught and cooked (what I will persist in calling) bullheads and been served them in a fine restaurant aboard a tourist ferry out of Cape Vincent, New York, and I've had channel cats (most of which are raised on farms now) in Midwestern restaurants and acquired live from a San Francisco supermarket for preparation and service at home.

Channel catfish may have been native to the upper Delaware River drainage, but for the most part they have been raised in Hacketts-town and introduced by the NJDEP into reservoirs, lakes, and ponds in every county in New Jersey. Both fingerlings (2 inches) and advanced fingerlings (5.5 inches) are stocked, and I count 111 water-bodies, ranging in size from diminutive Diamond Mill Pond in Mill-burn to our biggest reservoirs, in addition to the D&R Canal, on the NJDEP's list of warm- and cool-water stocking records for the nineties. This list is available simply by writing the Division of Fish,

Game and Wildlife's Bureau of Freshwater Fisheries at the Charles O. Hayford hatchery, but I've chosen a representative sample to get you started.

They are Hammonton Lake (Atlantic County); Scarlet Oak Pond (Bergen County); Mirror Lake (Burlington County); Haddon Lake (Camden County); Dennisville Lake (Cape May County); Bostwick Lake, Union Lake, and the Maurice River (Cumberland County); Weequahic Park Lake (Essex County); Greenwich Lake (Gloucester County); Woodcliff Lake (Hudson County); Amwell Lake and Spruce Run Reservoir (Hunterdon County); Farrington Lake and Hooks Creek Lake (Middlesex County); Rising Sun Lake and Holmdel Park Pond (Monmouth County); Lake Hopatcong (Morris and Sussex counties); Lake Carasaljo (Ocean County); Barbours Pond and Monksville Reservoir (Passaic County); Woodstown Lake (Salem County); Lake Ocquittunk (Sussex County); Milton Lake and Echo Lake (Union County); and Furnace Lake (Warren County).

Channel catfish can be distinguished from bullheads by their tails, which are deeply forked rather than squarish, but they also have sharper-looking noses and bluish gray backs with irregular dark blotches scattered about their sides. Older males may become darker and lose many of their spots, however. In view of the multitude of waterbodies in New Jersey where they have been successfully introduced, it serves little purpose to describe habitat except to say that they are bottom oriented but a bit more active in the water column than bullheads. Channel cats eat forage fish, clams, snails, and crustaceans. Shiners, crayfish, or worms are generally the anglers' choice of baits, but the catfish will take a spinner or plug, as well as "stink baits" such as strong cheese (Eddy and Underhill 1974). In nature, channel cats get to be about 14 inches at age five, and 20 inches by ages eight to ten. The biggest channel cat caught so far in New Jersey was 33 pounds 3 ounces and well over fifteen years of age (1978, Lake Hopatcong).

All bullheads are basically brown or black, so you can't rely strictly on coloration to tell them apart, although the brown bullhead generally has a yellow belly and light band at the base of its tail, and a related bullhead (yellow bullhead, also found in New Jersey but not as commonly) has white rather than brown barbels beneath its jaw. According to my notes from a graduate school ichthyology course, the black bullhead's body is shorter than that of the others, but the distinguishing characteristic is the pectoral spines. Those of the brown bullhead are stouter and strongly barbed, or serrated, on the posterior edge, whereas the black bullhead's are not. In either case, however, a bullhead's pectoral spines (and dorsal spines too, for that matter) should be respected. They are effortlessly capable of puncturing your skin and introducing a mild toxin that, depending on your sensitivity to such substances, can make your hand blow up into a hot, painful balloon. (I saw this happen to my friend Jim's dad once at his family's pond in Oak Ridge.)

All species of bullhead spawn in spring, building protective nests in shallow water. Browns are reported to make nests 6 inches deep in sand or mud, while blacks frequently make use of habitats ranging from hollow logs to muskrat burrows. Upon hatching and developing their full complement of body parts, the tiny catfish enter the wider world but are protected by one or (in the case of brown bullheads) both parents until they are about a half inch long. Guarded schools of these cute little jet black miniature adults are very easy to spot near the surface along margins of quiet waterbodies during May or June (depending upon that season's temperatures). As they grow older, they adopt a nocturnal habit, resting during the day in dense vegetation and feeding most actively around dawn and dusk. Taking rough averages of growth data for bullheads in various parts of the country as reported by Carlander (1969), I computed sizes-at-age (of both species merged) of 5 inches for age one, 10 for age five, and 12 for age eight or nine. Age-for-age, browns tend to have a size advantage over blacks, and there is no standing size record for a black bullhead in New Jersey. The record brown bullhead was caught in Fort Dix's Lake of the Woods in 1997 (4 pounds 8 ounces).

Bullheads are found, like sunnies and pickerel, throughout the state. Sometimes they may be too successful, becoming overcrowded and stunted, while in other situations their success at the expense of other species allows them to dominate the community but not shoot themselves in the foot in the process. One of those places may be Nomahegan Park in Cranford, where in the early and mid-fifties I caught many a nice bullhead on worms and not infrequently on a small Flatfish diving plug I was targeting for largemouth bass. I cannot attest to their ages, but a report of three "big brown bullhead" being caught in Nomahegan Park during early spring 1997 (*The Fisherman*, April 3, 1997) suggests that the bullhead population is still doing fine there. In situations where bullhead become a nuisance, however, they are difficult to eradicate due to their ability to bury themselves in the mud to avoid fish management toxicants (Carlander 1969). I discovered the same thing while participating in a pond reclamation and trout restocking project on Cranberry Pond near Amherst, Massachusetts, in 1968.

A sampling of other spots listed by the NJDEP (1994) and/or Perrone (1994) as offering good bullhead fishing includes Smithville Lake in Burlington County, Tuckahoe Impoundment #2 in Cape May County, Carnegie Lake and the D&R Canal in Mercer County, Farrington Lake in Middlesex County, Budd Lake in Morris County, Woodstown Lake in Salem County, the Raritan River in Somerset County, and Surprise Lake in Union County, but you can find dozens of other places close to you listed in either reference.

THE WHITE SUCKER.

This is the shortest of my stories about sport fishes of New Jersey. Shortest, because there is less to say about the family from personal, scientific, or number of representatives/family standpoints. But, if you fish enough, especially using worms during trout season in rivers like the north branch of the Raritan at Route 202, the downstream portions of the Flatbrook and Paulinskill rivers, and even the Delaware River, sooner or later you are going to catch one. The only ones listed in the NJDEP's inventory of freshwater fishes of New Jersey are the white, or "common," sucker (*Catostomus commersoni*), illustrated at the bottom of figure 3.13, showing suckers nosed into a low-head dam, and the creek chubsucker and quillback, which I mention in passing. The white sucker may take your worm or perhaps even salmon egg in April and give you a meritorious tussle on ultralight tackle; the others I doubt will cross your path. When I inked my figure of suckers, I was sure that another species lived in certain of our waterbodies, that being the northern—now called "shorthead"—redhorse (*Moxostoma macrolepidotum*), but not so, according to the state roster. Hav-

Fig. 3.13. Two suckers: shorthead (*above*) and a common one facing a dead end.

ing put forth the effort to include it in the illustration, however, I decided to briefly describe it in the event you should catch one in up-
state New York or someplace else.

Suckers have scaleless heads and thick fleshy lips that can be extended, enabling them to suck food off the bottom. They live in relatively clean rivers, feeding on midges, amphipods, fingernail clams,
and detritus (Carlander 1969), but move upstream during early spring
(i.e., potadromous behavior) to spawn in shallow gravelly riffles, often below dams, precluding them from getting any farther upstream.

The shorthead redhorse has been called the "cleanest," feeder
among suckers known to Eddy and Underhill (1974) by reason of its
eating more live organisms, like insect larvae, than rotting vegetation. It bites readily on bait, wet flies or small plugs and spinners, providing good sport in fast waters (which is why I thought some of those
I caught in the rapids immediately below the low-head dam on the
north branch might have been redhorses—didn't have a scientific
key on me years ago). Anyway, they are also reputed to be the best
tasting of the family, baked or smoked, given their white flaky meat.

The white sucker may be a foot and a half in length by the time it
may embrace one of your salmon eggs or farm-raised "red" worms—
a disappointing catch to the average trout fisherman but not a bad
fight. (Doug used to maintain a red worm "farm," a coffin-sized box
full of decaying leaves laced with bread crumbs and other stuff, and

spiked with mail-order brood stock, buried in the brush behind our house in Union — an act we all ridiculed him for but didn't hesitate to raid for bait.) White suckers may get as big as 20 inches (roughly 3 pounds), according to Carlander's (1969) data compilation, and the males will undergo quite striking changes in appearance with the arrival of spawning season. First, they develop "pearl organs" (a.k.a. "nuptial tubercles") on their fins. These are "little horny excrescences . . . that appear under the influence of hormonal secretions" (Lagler, Bardach, and Miller 1962). Then their color changes from dull gray to a vivid black with reddish markings, which they lose shortly after completing the spawning act. White suckers are also edible, but not as appetizing as redhorse.

CARPS AND MINNOWS.

This family, Cyprinidae, is the largest in the world, its members numbering more than 1,500 (Lippson and Lippson 1997). In North America alone, the AFS (1991) lists 240 species, 22 of which have been recorded in New Jersey, according to the NJDEP's list. It includes fishes as diverse as carp, goldfish, shiners, minnows by name, chubs, and dace. The seven I've illustrated here are the common carp (*Cyprinus carpio*), the goldfish (*Carassius auratus*), the golden shiner (*Notemigonus crysoleucas*), the common shiner (*Luxilus cornutus*), the fathead minnow (*Pimephales promelas*), the creek chub (*Semotilus atromaculatus*), and the blacknose dace (*Rhinichthys atratulus*). I made a last-minute decision to include mention of two other species, the grass carp (*Ctenopharyngodon idella*) and the "fallfish" (*Semotilus corporalis*), given that both are included on the NJDEP's list of freshwater species. Size records are maintained for grass carp, and fallfish have been cited as live bait for walleye.

There are only two that you would consciously try to catch, seven that might bite your hook anyway, three that are most commonly sold as bait, once you imitate with a streamer fly, and at least two you might imitate with a large spoon. The two that are actively pursued are the common and grass carps. Both are sought by rod and reel, and also by bow and arrow. The state rod-and-reel record for common carp is 47 pounds (1995, south branch of the Raritan River), but no one has yet landed a grass carp big enough to exceed the established

Fig. 3.14. Minnows: (*top to bottom right*) carp, blacknose dace, fathead minnow, and golden shiner; (*top to bottom left*) goldfish, creek chub, and common shiner.

minimum requirement of 30 pounds. In contrast, archers have gone into the record books with a common carp of 42 pounds 1 ounce (Delaware River 1987) and a grass carp weighing 49 pounds 3 ounces (D&R Canal 1996). Bowfishing has become as, or possibly more, popular a means of tackling carp than rod-and-reel fishing, although true carp anglers are a devoted lot (access "CarpNet" on the internet if you doubt me).

Carp have always been prized in Asia, where they originated, for their meat, roe, and ornamental value. They were among the first fish species to have been domesticated (i.e., pond raised). Aristotle mentioned carp in 350 B.C.E. and by 1227 European monks were raising them in ponds for food (Boyer 1995). By 1653 Izaak Walton was saying that "the Carp is the Queen of Rivers" and, quoting one Sir Richard Baker, that "Hops and Turkeys, Carps and Beer . . . Came into England all in a year!" (Walton [1653] 1936). This "Wonder Fish" was introduced with much excitement and aplomb into the United States in 1877, beginning with a shipment of 345 healthy specimens to New York Harbor that were then liberated into a pond in Boston, and via mass introductions or escapements in succeeding years quickly spread throughout the country (Laycock 1966). It wasn't too long, however, before they wore out their welcome among fisheries managers, most anglers, and some duck hunters (I'll come back to this later).

All members of the minnow family have heads without scales. Most (except carp and goldfish) have fins without spines, no teeth except for some rows of pharyngeal teeth in their throats, multiple ribs that make them excessively bony, and (among those big enough to eat) coarse flesh and a "muddy" flavor. Some (e.g., carp) crowd into shallow waters and deposit their eggs with much splashing, others take great pains to construct and/or preempt elaborate stone nests, and one deposits its eggs on the undersurfaces of various forms of cover. Minnows are most closely related to suckers, but their anal fins are not as far back and their lips are less suckerlike.

The carp is the biggest of the lot in North America. Its back is reddish brown or coppery, and its belly is lighter and more silvery. Some are fully scaled ("scale carp"), some partially scaled ("mirror carp"), and still others scaleless ("leather carp"). The scale carp is our most common variety. Those *with* scales have very big ones. Carp also have two barbels on each side of their upper jaws, and barbed spines at the front of their dorsal and anal fins.

Carp, which are very prolific (Eddy and Underhill, 1974, report that carp weighing 15 to 20 pounds may have over 2 million eggs), are spring spawners. They do this in waters frequently under 1 foot in depth on grounds characterized by Boyer (1995) as having a "muck bottom, shallow marginal grassy areas, intermittent inundation, exposure to air, presence of scattered bushes, and extensive growth of submerged and emergent vegetation." During the spawning act, which I have witnessed in Lake Lansing, Michigan, as well as some of New Jersey's larger rivers, they thrash about and make enough commotion to get your attention, muddying the waters in the process. They also muddy the waters when they feed, uprooting vegetation to get at the tender roots, or sucking up and spitting out bottom sediments from which they then selectively extract invertebrates of interest. This is one of the main reasons that they have become unpopular with many anglers, by increasing turbidity and supposedly damaging spawn or spawning habitat, and even with duck hunters. Laycock (1966) relays a story of Lake Koshkonong in Wisconsin, which was popular among hunters who came each year to slay canvasbacks which predictably stopped en route south to feed on an abundance of wild celery and pond weeds. Following the carp invasion, the hunter's quarry ceased answering the call.

Nonetheless, carp have their share of devotees since they are big, not too easy to catch, and (to some) good to eat when seasoned with an assortment of herbs and spices, then baked, broiled, or boiled (not pan-fried, according to Eddy and Underhill 1974). They can be caught on garden and meal worms, but dough balls and corn kernels are very popular baits. Many of my graduate school classmates went after carp religiously in East Lansing's (Mich.) stretch of the Red Cedar River using dough balls. Other baits include breakfast cereals and pet food biscuits glued to the shank of the hook, according to an account I scanned on the internet called "Surface Fishing for Carp" by someone named Peter, and "6-inch popped up corn baits," which you will need to check out in Manny Luftglass's (1996) book on carp fishing entitled *Gone Fishin' for Carp!*

Places to go include the ever popular midsection of the Delaware River, the Hackensack River (Bergen County), Passaic River (Essex and Passaic counties), Raritan River (Somerset County), Wooddale Park Pond (Bergen County), Crystal Lake (Burlington County), Blackwood Lake (Camden County), Carnegie Lake (Mercer County), Warinanco Park (Union County), and Woodstown Lake (Salem County).

Goldfish entered our waters more or less accidentally, but they are able to survive and reproduce. In the wild they generally assume olive or brownish hues, but I long ago caught vivid orange ones about 8 inches long where the feeder creek enters the pond alongside the skeet shooting range in Kenilworth. One bit a garden worm suspended from a bobber. They have spines on their dorsal fins but no barbels on their upper jaws, and they look more pugnacious than the carp with their more terminal mouths.

The golden shiners are the biggest and most deep bodied of the "shiners," growing as large as 12 inches but more typically, as seven- or eight-year-olds, about 8 inches long. They wear a translucent golden complexion but have also been described as green-bronze in color, changing to gold with brilliant orange-red fins during mating season (Lippson and Lippson 1997). Their lower jaw protrudes slightly beyond their upper, and their belly has a sharp keel. Golden shiners prefer clear weedy lakes with extensive shallows and scatter their eggs not only over filamentous algae and rooted aquatic vegetation, but even into centrarchid nests (Boyer 1995). They eat plankton, algae, and insects, but don't be surprised if you occasionally hook one on a small worm. Live golden shiners, which can be bought at bait stores and boat liveries near major lakes and reservoirs, make for excellent game-fish bait, being more active and less fragile than common shiners.

Common shiners, representing a different genus, are also sold as bait, and they reproduce in our streams as well. They are "silvery iridescent" and sport rosy pigmentation on their paired fins come spawning time (Eddy and Underhill 1974). Prior to spawning, males heap up small stones at the head of a riffle, or they may just commandeer a nest made by another species of minnow, such as the creek chub described later. While usually less than 6 inches in length, they too may grow to be capable of providing a nice (but brief) scrap on ultralight fly or spin fishing tackle using pieces of worms, tiny spinners, or wet flies.

In my experience on the streams of New Jersey, the minnow most apt to steal your worm or salmon egg, sometimes hitting a spinner too, is the creek chub. A creek chub may grow to 10 to 12 inches by the age of eight, providing some sport but mostly just vexation. In Walton's ([1653] 1936) account of the chub (he didn't specify creek chub), he notes that "the French esteem him so mean, as to call him *un vilain*; nevertheless he may be so dressed as to make him very good meat." The creek chub is normally silvery with a black band from head to tail and a black patch at the anterior base of its dorsal fin, but in mating season males develop coarse tubercles (pearl organs) on their heads and have multicolored sides. They also have a protractile premaxillary apparatus and silvery, often concealed barbel just before the ends of their maxillaries.

The male creek chub, using his nose to push or his mouth to carry pebbles upstream to the head of a riffle, creates a ridge parallel to the current that will serve as a nest, which he staunchly defends. A related species, the "fallfish" (*Semotilus corporalis*), makes a more ambitious nest of stones, measuring up to 5 feet in diameter and 2 to 3 feet high (Carlander 1969).

The fathead minnow is so named, I think, because its head looks fat during breeding season, given the presence of a large gray pad of nuptial tubercles on the top of its body from the back of the head to the front of the dorsal fin (not shown in my illustration). In this condition, such pearl organs also adorn its snout. The fathead has a protractile premaxillary but no barbels, and the adult has a horizontal bar midway up and extending the length of its dorsal fin, the first ray of which is thickened. They grow to lengths of 2 to 3 inches; most of those you buy in bait shops are about 1½ to 2 inches long. Fatheads are among the hardiest of the freshwater baitfishes, but what makes them more unique is their spawning habit and habitat. Specifically, they nest under boards, shingles, rocks, and even lily pads, and the male fertilizes the eggs *after* they are stuck to the undersurface of such cover by the female. The male then proceeds to guard them and stroke them using the pad on his back (Eddy and Underhill 1974).

One more species is illustrated because I wanted to give the reader some impression of what the "blacknose dace" streamer or "bucktail" fly is supposed to mimic if fished properly. The upper part of this dace's body is dark brown with black blotches, the lower white or silvery, and it has a faint but perceptible black lateral band stretched from its tail to its snout, traversing its eye. This line turns red during breeding season. The fly I use to imitate the blacknose dace, one that probably looks (to the trout) like either of two other species found in New Jersey (longnose dace and pearl dace), is a bucktail I know as the squirreltail mentioned earlier. Its body is made from the hairs of a brown squirrel, arranged from brown to black to white (top to bottom) and attached behind the eye of a size 12 hook with a shank about an inch long. The shank is wrapped in silver tinsel, and the hairs extend about a quarter inch beyond the end of the hook.

THE HERRINGS.

I elected to include "herrings" (from the family Clupeidae) in my freshwater fisheries section because, though the 7-pound state record American shad (*Alosa sapidissima*) was caught long ago (1967) in Great Bay, most shad and "river herrings" (alewife [*Alosa pseudoharengus*] and blueback herring [*Alosa aestivalis*]) are caught recreationally in fresh waters. Confounding this decision is the fact that all three split their allegiance between fresh and salt waters, and, because of familial ties, I was perforce obliged to discuss the Atlantic menhaden (*Brevoortia tyrannus*)—a nonsport baitfish species I associate primarily with salt water—in this chapter. Perhaps this is as it should be, forming a bridge to my saltwater chapter.

American shad and river herrings are anadromous, although the alewife is capable of living a landlocked existence. The Great Lakes are prime examples, but alewife have also been established in reservoirs, such as Merrill Creek, where they constitute an important prey base for large game fish. The menhaden, commonly known as "mossbunker" (or "bunker" for short), spawns and is caught commercially in the ocean, utilizing estuaries primarily as juveniles. Two other clupeids, threadfin and gizzard shad, also trade habitats.

It is because of their anadromy that shad and river herring populations have historically been so distressed by human enterprises.

56

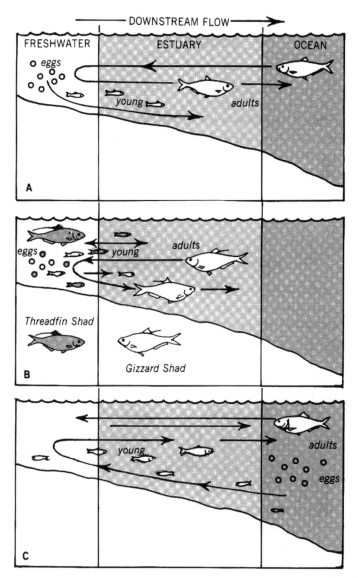

Fig. 3.15. Spawning patterns of the Alosids: Panel A represents American shad and river herrings, Panel C the Atlantic menhaden, and Panel B two other shads less common to our area. From Alice Jane Lippson and Robert L. Lippson, *Life in the Chesapeake Bay,* 2d ed., 119. Copyright © 1997 by Alice Jane Lippson and the Johns Hopkins University Press.

First and foremost among these have been impediments to finding ancestral spawning areas. Such impediments or complete blockades include dams, floodgates, and elevated culverts; dredge spoil or other forms of landfill in addition to road- and railbeds; mosquito control ditching; irrigation, muskrat, and waterfowl impoundments; and salt hay farming sites.

Zich's (1977) account of historical reductions in clupeid populations in New Jersey is the source of information to follow. It seems that in the late 1890s, the Delaware River boasted the best shad run on the East Coast. Not only the Delaware River run per se, but runs into most major tributaries from Trenton to Salem. That changed in the early 1900s due to physical blockages and/or water quality, but the same thing had happened much earlier on the Raritan River. It was 1872 when a U.S. Fisheries Commission said it feared that no spawning grounds were left due to discharge of "gas works" and rubber works waste in the New Brunswick area, and completion of the D&R Canal dam in the Raritan main stem below Branch Brook. "Gas works" were also implicated in destruction of the shad run in Blacks Creek, just below Trenton, as a result of "gas tar defilement." The Passaic and Hackensack rivers also supported shad runs, as did (in 1902) all of the creeks, streams, and coves of Barnegat Bay. Attempts to stock the Passaic River circa 1877 and the Raritan and Hackensack rivers in 1881 failed to restore successful runs. Finally, in 1916 the New Jersey Board of Fish and Game Commissioners' annual report decried the fact that pollution had destroyed great areas of shad spawning near Philadelphia — the tip of the iceberg of what was later to become known as the "pollution block" in the Delaware. In 1977, following four years of study, Zich reported the extinction of 9 herring runs and 19 shad runs in New Jersey, but confirmed that 133 clupeid spawning runs still materialized among 63 major drainage systems despite 83 barriers to optimal spawning (108 alewife, 24 blueback, and 1 shad run — the Delaware main stem — I'm not sure why he did not count the Hudson River shad run).

Fortunately, the Delaware River "pollution block" reference is now preceded by the word "former," several sources noting that the area has cleaned up its act with enforcement of the Clean Water Act (Bryant and Pennock 1988). Our interest and inventiveness have also allowed us to avail river herring of access to some ponds they were wont to spawn in. For example, a major electric company recently subsidized construction of a fish ladder to Sunset Lake in Bridgeton as a condition of its operating permit for one of its power plants, and with improving water quality, shad are again showing up in the Raritan River (forty-eight were counted in a new fish ladder during the 1996 season).

Let me digress, for a moment, back to the subject of "gas works." These plants produced gas from coal and/or oil using several fairly simple processes involving the use of steam, which I won't go into here, but which symbolized a very important breakthrough for home owners and industries in the late 1800s when all folks had for fuel was coal, oil, or wood. The technical term for the product was "manufactured gas," the plants being dubbed manufactured gas plants (MGPs).

The first gas product, initially bottled but later piped to customers, was used to light the streets of cities and suburbs across the country, allowing residents to safely walk the streets at night. The first plant in America was built on the south side of Baltimore harbor, but by the 1930s, practically every town and city in New Jersey from Cape May to Newton had at least one MGP, and the presence of an MGP was critical to growth of business, tourism, and industry in those towns. Every big town along the Erie Canal in New York State can also trace its growth and development to the moment a gas works was built. Though MGP products were eventually used for cooking and heating, the period from 1890 to 1940 is billed as "the gaslight era," and residents of towns across America, Europe, and Australia were proud to thus be part of the twentieth century. The downside was the masses of waste produced, like coal tar and byproducts of the purifying processes, which if not recycled as raw material for other businesses (e.g., asphalt) went largely into pits in the ground, sometimes finding their way into adjoining surface waters. The increasing use of electricity and later expansion of natural gas pipelines extinguished the MGP industry by the early fifties. Today, all that remains are the relics of foundations and an occasional gas holder (telescoping tanks within a frame reminiscent of an erector set), and these sites are gradually being sealed or cleaned to appropriate standards by today's utility companies. Having worked on a number of these sites in New

Fig. 3.16. American shad eyeing a "dart."

Jersey, New York, and California, I became fascinated with MGPs, but . . . back to clupeids.

American shad do spawn in both the Delaware and Hudson Rivers, and are quite a cause célèbre when they arrive each year in mid-April. "Hooked-on-the-Hudson," a shad festival cosponsored by the Hudson River Fishermen's Association, Hudson River Foundation, and Palisades Interstate Park Commission in Alpine, New Jersey, is one of those (Palisades Interstate Park Commission 1998). The Lambertville "shad festival," held the same time of year on the banks of the Delaware and growing more popular with each passing year, is another. If you arrive in Lambertville after 11:00 a.m., it may be too late to find parking near the center of town, but the "shad bakes" and other events and vendors can make the wait worthwhile. News can be accessed on the web site http://mgfx.com/fishing/regions/delaware/hotline, and shad fishing methods in the Delaware are available at www.acuteangling.com/shad.

The earliest recorded date for the arrival of shad in Lambertville is March 9, although their first appearance varies as a function of water temperature, which varies as a function of how cold the winter was, and how much snow fell and melted to be flushed downstream following the ski season. Early in the season, Delaware River shad "hot spots" (no pun intended) are the cooling water discharges from power plants at Duck Island (near Trenton), Martin's Creek, Pennsylvania (above Easton), and Portland, Pennsylvania (just below Columbia, N.J.) (Scholl 1994; Methot 1994; Johns 1994a).

In the Hudson, shad spawn 120 miles upriver of Manhattan on the sandbars between Kingston and Coxsackie. In the Delaware, shad may swim well into the headwaters to spawn, passing Hancock, New York, and ascending both the east and west branches downstream of two of New York City's main water supply reservoirs. Where each contingent chooses to stop and spawn depends on flow, temperature, and water quality.

I did a job for the New York State Department of Environmental Conservation in the early eighties that involved evaluating minimum in-stream flows needed to ensure maintenance of adequate fish habitats downstream of the Cannonsville Reservoir on the west branch of the Delaware River. That study was commissioned in part as a result of contemporary requirements for licensing dams, but the

Illus. 5. Shad fishing in the thermal discharge of a power plant on the Delaware River in 1998

importance of flow for downstream users had already been recognized by mandate of the Montague Act. The Montague Act, a negotiated settlement in the fifties between New York State, the city of New York (which owns the reservoirs), New Jersey, and Pennsylvania, mandated a formula for releasing enough water to help stay the worsening water quality problems in the lower Delaware, while still providing a sufficient supply of potable water for New York City residents. Shad and striped bass, both anadromous, were especially hurt by poor water quality in the Philadelphia area, but they are coming back again now that the quality of the river has improved.

From 1896 to 1990 the commercial landings of shad on the East Coast were decimated, most of the reduction occurring in the first quarter of the century, although the Delaware River population was experiencing somewhat of a comeback. Since 1972, when the Delaware River Anadromous Fishery Project initiated a program to estimate numbers of shad entering the river to spawn, their number increased to over three-quarter million in 1992 (Miller 1995). New Jersey Fish and Game mounted a hydro-acoustic fish counter onto the Route 202 bridge (just upriver from Lambertville) in 1992, counting 535,000 shad and about half a million each in 1995 and 1996 (Delaware River Shad Fishing Association "Hotline" referenced earlier). Sixty-one percent of the shad pass this point in April; the rest go by in the first nine days of May. A creel survey in 1995 estimated that Delaware shad anglers spent 300,000 hours over the course of 79,000 trips, catching 83,000 shad (Miller and Lupine 1996). Much of this activity takes place between Lambertville and Hancock, the uppermost 114 miles (i.e., above the "Water Gap") being designated a stretch of "Wild and Scenic River" (Scholl 1994). One particularly good spot to fish is the S-shaped bend called "Walpack Bend," just above the Water Gap, according to Scholl (1994) and my buddy Jim, who has done a lot of shad fishing.

Males generally arrive first on the spawning and sport fishing grounds, averaging 2 to 6 pounds, while the larger 8- to 9-pound females follow (Geiser 1969). Shad are silvery with a dark green-blue back, a saw-edged belly, soft-rayed anal and dorsal fins, loose scales, and a fragile mouth that needs to be reckoned with after you've hooked one. Most fishing is done with light tackle using a "shad dart," which looks like a painted BB split shot weighing an eighth to a quarter ounce with a short bucktail, cast cross-stream and to coil downstream while you retrieve it in slow, consistent pulls of about 1 foot each (Geiser 1969).

Unlike salmon, shad head downstream and out to sea immediately after spawning and may return some two to five more seasons to repeat the act (Miller 1995). Juveniles stay in the river until fall, emigrating when they are between 1½ and 4½ inches in length (Bigelow and Schroeder 1953). Shad roe is a delicacy, and their meat is also excellent, though very difficult to filet given the fact that shad have three rows of "pin bones" versus the usual single row of other bony fishes.

Alewife and blueback are similar in appearance to small shad, but they have smaller mouths and shorter upper jaws. Like shad, both are noted for their sharp ventral keels and are commonly called "saw-bellies." Alewives have bigger eyes than blueback, and blueback have a bluer back. Both spawn in the Delaware main stem past Trenton, but are most noted for their runs into tributaries and coastal ponds still having an unoccluded link with the Delaware or the sea. The alewife likes best to spawn in ponds but will settle for sluggish stretches of streams, whereas the blueback prefers faster water and may settle for habitat short of ponds. Both species spawn in spring, and at an average size of 10 inches make worthy sport where they throng in and below spawning sites. According to Geiser (1969), the "Wreck Pond Pipe" off the Metedeconk River and the Deal Lake Flume into the ocean were traditional favorite places to fish using something as rudimentary as a small, bare gold hook. On the Delaware side, the Maurice River always provides good river herring fishing in April and May. Spent alewives head back to sea soon after spawning, while spent blueback may linger through summer in fresh or moderately salty water (Bigelow and Schroeder 1953). Juveniles leave their natal rivers and swim toward the sea from early summer through early fall, returning three years later to fulfill their destiny.

Unlike shad and river herrings, menhaden are not a target of sport fishermen, but they are directly or indirectly linked to the angler's success. As live or dead bait, or as minced "chum," they are directly linked to fishing for gamefishlike striped bass, bluefish, and fluke.

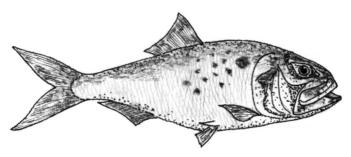

Fig. 3.17. Menhaden, or "bunker."

The bodysurfer part of me finds chumming to be distasteful, smelly, and irritating, but this gruel does attract fish and increases the likelihood of making a good day's catch. Indirectly, menhaden are a staple in the diet of every predator in the sea—from whales, to cods, to tunas and blues (their worst enemy)—and maintenance of their population is considered fundamental to the success of the sport angler's target species. And there's the rub, insofar as purse seining for menhaden is third most important to New Jersey's commercial fisheries, numbers one and two being clam dredging and otter trawling. Exacerbating the conflict are the occasional news releases about commercial rogues who do not obey the rules, and the fact that commercial menhaden fishermen and recreational anglers are always bumping into each other while fishing the same waters off our coast where they are entitled to fish.

Mossbunker have always been a very important commercial species, perhaps *the* most important in the United States (Hall 1995). They are "industrial" fish, their oil being used in the manufacture of paints, cosmetics, and more recently margarine (approved by the FDA in 1997), and their meat being an excellent protein supplement in feed for hogs and poultry. Depletion of menhaden stocks along the East Coast in the seventies and early eighties led the Atlantic States Marine Fisheries Commission (ASMFC) to target menhaden as the first species to be in need of a Fishery Management Plan (FMP) (Hall 1995). That plan being implemented and purse seining banned in Delaware Bay, New Jersey's commercial menhaden landings have rebounded from 1.6 million pounds in 1983 to 36 to 38 million pounds from 1994 through 1997. Still, the controversy continues, with commercial fishermen claiming that more can be taken, and sport anglers insisting that commercial fishermen are being given far more than the population can withstand. As this book went to press, a bill to push purse seiners further offshore was before the New Jersey Senate.

Menhaden are toothless, they have scaleless heads, and the rear margins of their body scales are nearly vertical as opposed to the rounded margins on shad and herring scales. The inside of the mouth is sometimes shared with a parasitic isopod (a bug similar to the "sow bugs" in your garden or under logs in your woodpile), a relationship blamable for their being labeled "bugfish" (Geiser 1969).

The Waterbodies

Noted earlier was the fact that New Jersey has many thousands of miles of running waters and tens of thousands of acres of standing waters. Places the public can fish these waters are listed in the NJDEP's (1994) *Places to Fish—List of NJ Lakes, Ponds, Reservoirs, and Streams Open to Public Angling*, also accessible via www.state.nj.us/dep/fgw/ fishplc. Luftglass and Bern (1998) also provide synoptic profiles of many freshwater lakes, streams, and reservoirs in their book *Gone Fishin': The 100 Best Spots in New Jersey*.

All of these waterbodies are contained within five major drainage basins, the largest being the Delaware River Basin, which drains 3,000 mi^2; the next being the Atlantic Coastal Basin, draining 2,000 mi^2; the next being the Passaic and Hackensack Basin (1,200 mi^2); followed by the Raritan River Basin (1,100 mi^2); and the last being the Wallkill River Basin (210 mi^2), which oddly enough drains a portion of Sussex County near Hamburg from south to north, entering the Hudson River many miles north of our border. These basins contain ninety-five principal "watersheds," which the state has arranged into twenty major "watershed management areas" of roughly equal size (NJDEP 1998a). Within each are a wide variety of habitats ranging from "wild trout" habitat in the north to "Pinelands" streams in the south. The former get their water from rainwater and springs near the top of the hills in their watersheds, the latter from acidic groundwater springs of the Pine Barrens (although more and more water pouring into these streams comes from overland runoff and sewers

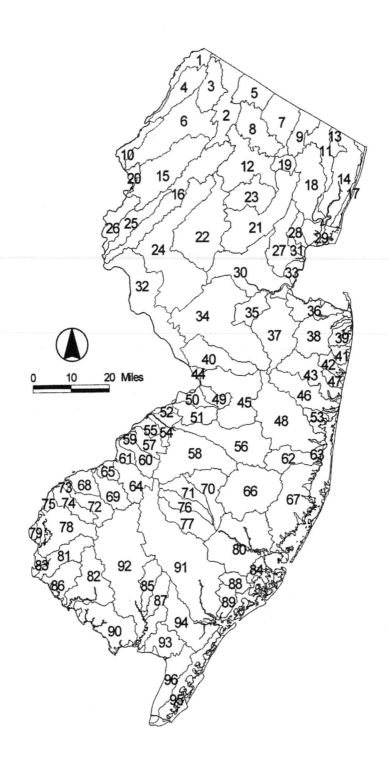

created by paving and sewerage associated with development). The type of fish found in these watersheds is a partial function of the area's natural geology and topography, modified by human activity.

From a geological standpoint, New Jersey is divided into two main "provinces": the Appalachian Province and the Coastal Province. The Appalachian Province extends from our coastal plain to the

---

Fig. 3.18. (*opposite*) Principal watersheds of New Jersey. From NJDEP (1998a).

1. Shimmers Brook	33. Woodbridge River	65. Woodbury Creek
2. Wallkill River	34. Millstone River	66. Wading River
3. Papakating Creek	35. Lawrence Brook	67. Forked River
4. Flat Brook	36. Matawan Creek	68. Repaupo River
5. Pochuck Creek	37. South River	69. Mantua Creek
6. Paulins Kill	38. Navesink River	70. Batsto River
7. Wanaque River	39. Shrewsbury River	71. Atsion Creek
8. Pequannock River	40. Assunpink River	72. Raccoon Creek
9. Ramapo River	41. Whale Pond Brook	73. Maple Swamp
10. Vancampens Brook	42. Shark River	74. Oldmans Creek
11. Saddle River	43. Manasquan River	75. Whooping Creek
12. Rockaway River	44. Duck Creek	76. Mechesactauxin Creek
13. Pascack Creek	45. Crosswicks Creek	77. Nescochaque Creek
14. Hackensack River	46. Metedeconk River	78. Salem Creek
15. Pequest River	47. Wreck Pond Brook	79. Miles Creek
16. Musconetcong River	48. Toms River	80. Mullica River
17. Hudson River	49. Black Creek	81. Alloways Creek
18. Lower Passaic River	50. Crafts Creek	82. Cohansey River
19. Pompton River	51. Assisicunk Creek	83. Mill Creek
20. Delawanna Creek	52. Mill Creek	84. Doughty Creek
21. Upper Passaic River	53. Kettle Creek	85. Manantico Creek
22. N. Br. Raritan River	54. Rancocas Creek	86. Stow Creek
23. Whippany River	55. Pompeston Creek	87. Manumuskin Creek
24. S Br. Raritan River	56. N. Br. Rancocas Creek	88. Absecon Creek
25. Pohatcong Creek	57. Pennsauken Creek	89. Patcong Creek
26. Lopatcong Creek	58. S. Br. Rancocas Creek	90. Dividing Creek
27. Rahway River	59. Baldwin Run	91. Great Egg Harbor River
28. Elizabeth River	60. Coopers Creek	92. Maurice River
29. Elizabeth Channel	61. Newton Creek	93. Dennis Creek
30. Lower Raritan River	62. Cedar Creek	94. Tuckahoe River
31. Moses Creek	63. Sloop Creek	95. Cape May Atlantic Coast
32. Lockatong Creek	64. Big Timber Creek	96. Delaware Bay Coastal

Mississippi River lowlands, while the coastal plain stretches from Massachusetts Bay to the Gulf of Mexico. In New Jersey, the boundary between these two provinces is the "fall line" extending from Trenton to New Brunswick.

Within the Appalachian Province, proceeding due northwest in bands parallel to the fall line, one first crosses the "Piedmont Plateau." This is the largest band, its northwest border going from just north of Milford, on the Delaware, through Pottersville, then Morristown, Pompton, and ultimately to Bergen on the New York border. The Piedmont encompasses 20 percent of the state, including all of Essex, Hudson, and Union counties; most of Bergen, Hunterdon, and Somerset counties; about half of Mercer County; and about a third of Middlesex, Morris, and Passaic counties. Within that area flow the Hackensack, Passaic, Rahway, Millstone, and Raritan rivers (including all of its north branch plus half of its south branch). The Piedmont lies, geologically, in an area of metamorphosed (changed) igneous rock (e.g., granite, formed by cooling and solidification of hot liquid rock material) and sedimentary rock (e.g., shale, or compacted clay or mud from weathered rocks). On top of this, waterbodies of much of the Appalachian Province have been blanketed with deposits of sand and gravel called "alluvium."

The upside of the next—"Highlands"—band is defined by a line starting just south of Brainards on the Delaware, through Belvidere, Springdale, Lafayette, and Sussex, ending at the border near Unionville, New York. The Highlands contain most of nothern New Jersey's large water supply reservoirs; Lakes Hopatcong, Mohawk, Musconetcong, and Waywayanda; Budd and Highland Lakes; the upper halves of the south branches of the Raritan and Black rivers; and the Musconetcong, Pequest and Wallkill rivers. Highly metamorphosed crystallized rocks (gneiss, quartzite) and limestone (e.g., marble, formed of weathered and compacted shells of marine origin deposited when the sea covered the state), are the foundation of the area's geology. Localized iron and zinc lodes are found, respectively, around Highland Lakes and Franklin.

The uppermost band, the Kittatinny Valley and Ridge, includes New Jersey's highest point (High Point, 1,804 feet above sea level) and the Paulinskill and Flatbrook rivers. Sandstone (consolidated sand) is what Kittatinny Mountain is made of, while limestone forms the bed of the Kittatinny Valley. Shale outcroppings can be seen, and boulders, rocks, and pebbles, scoured by the great glacier or eroded from hills, dot the surface of the Kittatinny's waterbodies.

The coastal plain is formed mostly of beds of clay, sand, and gravel. Thick beds of clay have historically been mined from Camden through Cape May counties, but while clayey material is found along the Manasquan and Shark rivers, sand assumes prominence in more southern coastal streams like those emanating from the Pine Barrens. Surface waterbodies originating in the Pine Barrens are very acidic, reflecting the nature of the Kirkwood-Cohansey groundwater, which feeds them through springs. That level of acidity is one of the main factors distinguishing these waterbodies from all other streams. The Coastal Province extends some hundred miles out into the ocean where, upon dropping abruptly to the abyss, the province and continental shelf terminate.

As you will see later, the submarine geology off our coast also creates special fishing areas for New Jersey anglers. For those more interested in this fascinating subject, I encourage you to read Wilber and Johnson's classic publication *The Geology of New Jersey* (NJDCD 1940), which was my source of the foregoing synopsis. Now follows a characterization of the waterbodies running through or sitting in these provinces, from a fisheries standpoint.

RUNNING WATERS.

Running waters can be classified in several ways, starting with a system based on stream "order," In that system, the first little feeder streams off the top of the watershed are called "first-order streams" (such as Stony Brook in Stokes State Forest); streams created by the merger of two of these types form a "second-order stream"; and so forth. All things being equal, lower-order streams (i.e., first or second) would be expected to have fewer species of fish, and smaller ones, than higher orders. In northern New Jersey, an angler can pretty well assume that small trout will be the main (or only) sport fish to be found in a first-order stream, whereas smallish pickerel may be the main fare in first-order streams in the southern part of the state.

Lagler, Bardach, and Miller (1962) describe a progression of characteristics associated with (unstocked) trout, pike, bass, catfish, and carp ecosystems. Attributes of running waters at opposite ends of this spectrum (trout vs. carp) are as follows: (1) V- vs. U-shaped valley, (2) high vs. low gradient (therefore current velocity), (3) clear vs. cloudy water (turbidity), (4) rock and gravel vs. sand and muck bottom, (5) little vs. much organic material (e.g., from decomposing stuff or natural fertilizer), (6) cold vs. warm, (7) competition between members of the same species vs. among different species, and (8) few (if any) vs. several predators. This progression of conditions characteristically occurs as you pass through different "reaches" of a stream. In the classical progression, upper reaches are steepest, with swift current, ledges, boulders, and some pools. Middle reaches have equal numbers of pools and riffles, and some siltation, but floodwaters are confined to the channel. Lower reaches have no riffles, pools elongate, siltation is heavy, and floodwaters overflow the banks.

Reschke (1990) defines seven categories of streams in New York State: rocky headwater streams (like you'd find in Stokes Forest), marshy headwater streams (like those of the upper Rockaway), mid-reach streams (like the Paulinskill near Blairstown), main channel streams (like the Pompton River along Route 23), backwater sloughs (like some around the mid-Passaic River), intermittent streams (that only exist after rain or when snow melts), and coastal plain streams (like perhaps our Toms River). Many decades ago, one of my college professors produced a mimeographed handout of a Maryland stream classification system. In that system, ten kinds of streams were defined according to physical and water quality characteristics, each named after an associated fish species (starting with "dace trickle" and ending with "carp stream"). You might also read or hear reference to "freestone" versus "limestone" streams. Those terms are more commonly used in Pennsylvania, but Trout Unlimited (1975) invokes them to illustrate the Pequest River. Eastby (1994a) uses the term limestone in reference to the headwaters of the Paulinskill River above Lafayette, and Peinecke (1994a) employs the term freestone in depicting the upper part of the "Blewett Tract" (a fly-fishing-only portion of the Big Flatbrook below Route 206 in Sussex County). One more term, "meadow stream" (commonly associated with rivers in parts of states like California, Colorado, and Montana), is used to describe the portion of the Paulinskill downstream of Lafayette and places on the headwaters of the Pequest (Eastby 1994b) and Dark Moon Brook in Sussex County (Peinecke 1994b). Finally, the NJDEP classifies running waters FW1 (exceptional waters that must be protected from degradation), PL (Pinelands waters), and FW2 (all other fresh waters), adding to that whether or not they are "Trout Production Waters" (those that can support spawning and nursery functions), "Trout Maintenance Waters" (those in which trout can survive year round), or "Nontrout Waters."

Based on elements of all of these, but recognizing the tremendous variability in conditions within even a single mile or two of the same stream in New Jersey, I decided to create some categories (or "types" of waterbodies) that might have more pragmatic value to anglers throughout the state. I started by reflecting on characteristics of all of the places I've fished over the years, what I caught, and why I went there — to catch a particular species, to fly-, spin-, or baitfish; to be close to home; to be away from other people; or simply for ease of parking, given acceptability of other criteria. After much hand-wringing, I defined six types of streams or stretches of streams: (1) *Finite Development/Trout,* (2) *Trout/Smallmouth Bass,* (3) *Mixed Bag,* (4) *Mid-Jersey/Coastal,* (5) *Pinelands,* and (6) *Wild Trout.* Although basin variability still makes uniform portrayal a thorny matter, I tried to characterize each type according to approximate measures of width, depth, current velocity (during normal flow conditions), temperature, pH, turbidity, substrate, pools and riffles, land use development, and principal target or nontarget species. Most of these terms have been discussed before, or are self-explanatory, but let me just mention pH (7 being neutral, less than 7 being acidic, greater than 7 being basic, or alkaline). As noted earlier, there is a big difference between the pH of Pinelands waters and any others, although some first-order streams in the hills of North Jersey may be on the acidic side. The pH is a variable that can affect the ecology of running waters, and, as explained in chapter 5, it is one of those parameters on which the state has placed both upper and lower limits (water quality criteria). The lower limit only comes into play with pollutant spills or discharges, but the upper limit can change for the worse on a chronic and incremental basis with watershed development. To protect natural ecological communities, the upper limit on pH for Pinelands waters has been set at 5.5, whereas other waters have an upper limit of 8.5.

The sections to follow provide brief portraits of aquatic communities associated with each of my six categories, and examples of places I feel (from visual observations, not measurements) fit my physical descriptions. Two waterbodies I couldn't coerce into any of these categories, however, are the freshwater portions of the Delaware River above Trenton (the lower portion is described in chapter 4, estuarine waters) and the D&R Canal (although I was tempted to force-feed this one into "mixed bag"). The Delaware is wider than the other streams or rivers included in my classification scheme, and its fisheries range from shad (which may go all the way into New York on their spawning migrations), to striped bass (which are found in fresh waters during spring and, apparently, even into summer—see next chapter), to catfish/carp, largemouth and smallmouth bass, muskies, and walleye, to trout. Its waters are quite clean, and there are a variety of habitat types you can access from either shore or boat, using one of the many ramps provided by the state along Route 29 in Mercer, Hunterdon, and Warren counties. The Delaware River starts above Port Jervis at the confluence of its east and west branches, then flows (as an official "Wild and Scenic River") down through the Delaware Water Gap at routes 46 and 80. Above the Water Gap, the Delaware sports a diverse assemblage of substrates suitable for colonization by benthic invertebrates like insect larvae. In the late nineties, scientists from the Philadelphia Academy of Natural Sciences found about 150 species of invertebrates, mostly insect larvae, from cobble and pebble riffles along the upper Delaware and its tributaries (ANSP 1998).

In addition to the shad web site referenced earlier, the Delaware Division of Fish and Wildlife (Dover, Del.) also maintains a web site called the "Delaware Fishing Report" at www.dnrec.state.de.us/fw/fishrept.htm. Further information about the Delaware River and many other streams included in this chapter is also provided in Perrone's (1994a) book entitled *Discovering and Exploring New Jersey's Fishing Streams and the Delaware River*.

The D&R Canal is another story. Had the canal been proposed after 1969, chances are that it would not have advanced beyond the environmental impact statement process, thus denying New Jersey some rich fishing opportunities. The canal runs parallel to the Delaware River from Lambertville to Trenton along Route 29, then heads northeast along Route 1 through Lawrenceville and Princeton, and on up through Somerset County to its confluence with the Raritan River. As a canal, built for horse-drawn or poled barges, the D&R has a rectangular profile lacking features like pools, riffles, or meanders characteristic of flowing waterbodies profiled below. However, it is stocked with trout during spring, while also accommodating species like pike, largemouth bass, and carp. In Somerset County a good part of the canal is undeveloped (D&R Canal State Park), but shoreline access is abundant throughout most of its length, and it is an excellent choice for luring someone to fishing in a picniclike setting.

My *Finite Development/Trout* streams or stream sections are all contained within the area bounded by routes 78 (east-west) and 287 (north-south). In these places you can feel as if you're in the woods and, except for opening day and the next few restocking weeks, not snagging or being hooked by another angler. They are for the angler who likes wading the stream rather than standing on a bridge or sitting on the bank, but some (Pequest River along Route 46, Black River in Hacklebarney State Park) are treacherous to wade during high (fast) water or when algae begin to make the rocks slippery in late spring. They are generally not the sorts of places you should expect young kids to get hooked on the sport of fishing, for they can be frustrating to the beginner (more snags or tangles in the branches than fish on the line) and shaky (slips, slides, and soakings) to downright perilous for a bantamweight.

All waterbodies of this type are classified by the NJDEP as being either trout maintenance or trout production waters, and several stretches are regulated as "Year-Round Trout Conservation Areas," "Seasonal Trout Conservation Areas," "Fly Fishing Areas," or "No-Kill Areas" (NJDEP 1998b). The fish community is comprised of trout, dace, and creek chubs, but some common shiners, sunnies, or even a pickerel or two may show up in "meadows" areas with slower current and enough refuge. Still, they are mostly known and fished

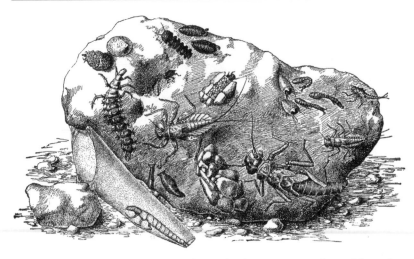

Fig. 3.19. Common invertebrates of a swift, clean stream. Adapted from Ruttner (1963, 236). Courtesy of the University of Toronto Press.

for trout. The rest of the food web includes attached algae, freshwater "scuds" (or shrimplike little amphipods), small freshwater clams, and a nice assortment of insect larvae and other critters that like living on rocks in clean water. Onshore, I've run into timber rattlers along the Flatbrook, and though I've never encountered a copperhead despite probably a hundred visits, the Black River is supposed to be within their range. As far as the rattler goes—while you don't want to get bitten, bear in mind that it is illegal to *harm or harass* them. The timber rattlesnake is on the New Jersey endangered species list.

Specific attributes of waterbodies of this type are varied. From a scientific or fly-fishing purist's standpoint, one could legitimately question inclusion of the upper reaches of the Big Flatbrook in the same category as the upper Rockaway. But, in trying to paint a portrait of places where the average angler might derive similar overall gratification, and broadening the geographic range of possible opportunities as much as reasonably possible, the following diagnostics and examples make sense to me:

- Average Width: < 30 feet
- Average Depth: < 2½ feet
- Current Velocity: high to moderate

- Temperature: cool to cold
- pH: usually 7–8; some excursions above the 8.5 criterion below towns
- Turbidity: low
- Substrate: rocks, cobble, stones, pebbles, gravel (silt in pools of the Rockaway)
- Pool/Riffle Ratio: $\geq 2:1$
- Land Use Development : light; wooded banks
- Principal Target Species: trout
- Examples: Big and Little Flatbrook rivers upstream of Walpack (Sussex County; note that the pH of the very upper portion of the Big Flatbrook in Stokes Forest may be less than 7, slightly on the acidic side); Musconetcong River in Stevens State Park and again between Hackettstown and Asbury (Warren County); Pequest River along Route 46 near Oxford (Warren County); south branch of the Raritan River between Long Valley and Middle Valley (Morris County), and again in Ken Lockwood Gorge between Hoffman's Crossing and High Bridge (Hunterdon County); Black River from Route 24 to about 1 mile downstream, and again in Hacklebarney State Park (Morris County); Rockaway River in the Newark water supply area along Berkshire Valley Road south of Oak Ridge, and between routes 15 and 80 (Berkshire Valley Wildlife Management Area in Passaic County); Wanaque River above Wanaque Reservoir (Passaic County)

Trout/Smallmouth Bass waters are streams or portions of streams situated farther downstream and generally in more developed areas, be they suburban or agricultural. They can be either waded or fished from numerous places along the shore, since many have a more open shoreline and/or deeper holes cutting right along the bank. Though rocky sections exist, for the most part the substrate is easier to tread upon, and you won't have to watch the kids quite as rigorously (partly because there is enough room for you to be near them). All of these places are in the northern half of the state, and all are stocked during trout season (trout maintenance waters).

The substrate and flow are such that the lower end of the food

Illus. 6. The Musconetcong River a few miles upstream of Asbury in Warren County

chain includes fewer species of may- or stone-fly nymphs, but caddis flies are present along with growing numbers of midge larvae, damsel-fly nymphs, and crayfish. The fish community includes trout, and that's what most folks go after during spring, but suckers, shiners, sunnies, rock bass, and, more importantly, smallmouth bass also enter the mix. The habitat is usually good enough for trout to find pools in which to hold over during summer.

Most fishing is done with artificial lures (spinners or small spoons) or bait (salmon eggs, worms, grubs, and fathead minnows), but the fly fisherman can still count on running into black gnat, caddis, or even some mayfly hatches. Here are some general characteristics and examples:

- Average Width: > 30 feet
- Average Depth: > 2½ feet

- Current Velocity: moderate
- Temperature: cool to warm
- pH: > 7.5; some places experience regular excursions above 8.5
- Turbidity: moderate to low
- Substrate: rocks, gravel, silt in pools
- Pool/Riffle Ratio: > 2
- Land Use Development: varies; still with relatively vegetated shorelines
- Principal Target Species: trout in season, but smallmouth bass are also popular, and pickerel may be found
- Examples: Rockaway River downstream of Dover (Morris County); south branch of the Raritan River along Route 31 (Hunterdon County), and again at Neshanic Station (Somerset County); north branch of the Raritan River from just downstream of Route 202 to North Branch (Somerset County); Paulinskill River at Blairstown (Warren County); Musconetcong River below Asbury (Warren County); Ramapo River (Bergen County)

Flowing waters in my *Mixed Bag* category are the kinds of waterbodies found in heavily developed watersheds that provide opportunities for a variety of cool- and warm-water species when you are limited by time or lack of wheels. They are typically wide and somewhat sluggish compared to streams in the northwest part of the state, but there are exceptions relative to width (the Rahway River in Union County being one). Most have experienced a variety of past municipal, commercial, or industrial insults as well.

Except for dragonfly nymphs (large things that look like compressed brown bumblebees before they climb up rocks, sticks, or bulkheads to molt and emerge as the four-winged colorful insects we know and love), these waterbodies don't support much of an insect community. Burrowing wormlike organisms, crayfish, and snails are more common, but there is enough live and dead organic material to produce lots of fish. These are "Nontrout Waters," but some receive plantings before opening day. Members of the pickerel family are more common, as well as catfish and carp, but sunnies, crappies, and even

Illus. 7. The Paulinskill River at Blairstown

some largemouth bass may be found. Live bait and lures are the order of the day in waterbodies of this type, and bait-casting equipment
can be used sometimes just as effectively as spinning gear. Characteristics and examples follow:

- Average Width: > 40 feet
- Average Depth: > 3 feet
- Current Velocity: low to moderate
- Temperature: warm
- pH: ≈ 8 with frequent excursions above 8.5
- Turbidity: moderate to high
- Substrate: mud, silt
- Pool/Riffle Ratio: > 10
- Land Use Development: moderate to heavy, but some
  stretches still wooded
- Principal Target Species: trout during stocking season in
  some, others stocked with pike or tiger muskies, but more nat-
  urally a catfish/carp type of stretch

- Examples: Hackensack River above Hackensack (Bergen County); Rahway River (Union County, off Exit 136 of the Garden State Parkway near Cranford and Winfield); Crosswicks Creek (Burlington County; tiger muskies stocked); Passaic River (Bergen County; tiger muskies and channel catfish stocked); Raritan River at Piscataway (Middlesex County); Pompton River (Passaic County; tiger muskies stocked); Millstone River (Somerset County; northern pike stocked); Rancocas Creek main stem (Burlington County; tiger muskies stocked; headwaters start in Lebanon State Park Pinelands area and are acidic, even though they now normally exceed the upper Pinelands pH criterion of 5.5)

The main thing that my *Mid-Jersey/Coastal* streams have in common is that they are all in Monmouth County or northern Ocean County, and all receive plantings of hatchery trout during early spring. Most are pretty tough, physically, to fish, and postings by pri-

Illus. 8. The Pompton River in Wayne

vate landowners are fairly commonplace. Suburban sprawl is the main threat to these watersheds. In some watersheds, like the upper Shark River, construction activity causes pH depression by exposing sulfuric acid–producing soils to air and water. All still offer some nice fishing opportunities for residents of the North Jersey shore.

Given the variability in habitat conditions, I need to get ahead of myself here in describing ecological conditions relative to waterbody examples. While most of these waterbodies normally contain low to moderate flows and comparably low to moderate levels of turbidity (i.e., they are fairly clear), one — the Manasquan River — stands apart. It is generally more turbid, and — draining 81 square miles — it carries a lot of water on its way past Brielle and into the ocean. Silty pools and runs outnumber pebbly riffles, but it has lots of meanders, undercut banks, and fallen timber to create habitat variety. Its upper reaches are classified as trout maintenance waters, and it is stocked with nice brownies and rainbows. When overland runoff is low for sustained periods, spinners and lures work well, but earthworms, mealworms, fathead minnows, and salmon eggs are generally the angler's best bets. Mosquito and blackfly hatches occur, but I don't think of this as a fly-fishing stream.

On the other hand, clearer streams like the mid to upper stretches of Toms River and Hockhockson Brook, with their complement of dace, caddis flies, and black gnats, usually offer good streamer and sometimes worthy nymph or dry fly-fishing (Lanzim 1994; Boa 1994). You can always rely on spinners to imitate dace, or earthworms to be themselves. I have only fished these streams till about June, but the authors just cited report that both waterbodies can hold some trout over summer.

To maintain consistency in format, here is my attempt at a global characterization:

- Average Width: < 30 feet
- Average Depth: < 3 feet
- Current Velocity: varies; all watersheds quickly affected by heavy rainfall
- Temperature: warm to cool

- pH: with one exception, > 7.5
- Turbidity: generally low to moderate except during heavy rains
- Substrate: varies; gravel in upper Monmouth County, then clayey mud and pebbles, to sand in Ocean County; pools and lazy meanders usually contain silt
- Pool/Riffle Ratio: varies depending on waterbody
- Land Use Development: moderate to heavy, but many fishing stretches still wooded
- Principal Target Species: trout during spring, maybe some pickerel in other seasons (tidal portions of some have white perch action in winter, covered in next chapter)
- Examples: Yellow and Hockhockson brooks and the Manasquan River from Allaire State Park upstream to Route 9 (Monmouth County); Metedeconk River upstream of Garden State Parkway (Ocean County); north branch of Toms River

Illus. 9. The Manasquan River just upstream of Allaire State Park at Route 547

upstream of Pleasant Plains (Ocean County; upper parts are acidic and rated "Pinelands—Trout Maintenance," lower reaches designated "FW2—Nontrout")

With the exception of two borderline cases (Rancocas Creek and Toms River), which, after a considerable amount of mulling over, I chose to include in other categories, *Pinelands/Pickerel* waterbodies were the easiest to define. All are classified "Pinelands—Nontrout"; all derive their flow from the acidic ground- and decomposing humic material–laced waters of the Pine Barrens, flowing east to the ocean; all have sandy substrates and clear, tea-colored water; and all are basically pickerel streams. Another fact—all are *very* popular among canoeists. I once led my pack of Webelos Scouts on a weekend camping and canoeing expedition down the Wading River, and I would not have taken much delight in being in the middle of the river—fishing—when they came down! Bank and weekday fishing are a different story. James and Margaret Cawley's book, *Exploring the Little Rivers of New Jersey*, provides a colorful and historic view of these and other streams from the canoeist's vantage point, including the great bog-iron works, charcoal producers, and mills.

With a watershed of 561 square miles, the Mullica is the Pine Barrens' primary drainage system (NJDEP 1998a). Eighty percent of the Mullica River and its tributaries flow through state parks or forests, and much of their lengths are included in New Jersey's Wild and Scenic Rivers system. Endemic to these parts are the endangered pine snake, Pine Barrens tree frog, black banded sunfish (looking like its name), and the occasional timber rattler. I found (with a simple search) three web sites devoted to Pinelands waterbodies, each of which seemed pretty accurate to me: (1) www.bpbasecamp.com/wilderness/nj_pinebarrens, (2) www.igc.apc.org/maurice-river, and (3) http://inlet.geol.sc.edu/MUL/site. Here is my synoptic list of further attributes and examples of Pinelands waters:

- Average Width: < 30 feet
- Average Depth: < 3 feet
- Current Velocity: moderate to high
- Temperature: cool in woods, but may get to 75°F downstream

- pH: < 5; some places exceed the Pinelands upper criterion of 5.5
- Turbidity: low (but highly colored water)
- Substrate: sand
- Pool/Riffle Ratio: > 1 (also lots of "chutes" and "meanders")
- Land Use Development: very light (so far)
- Principal Target Species: pickerel
- Examples: Mullica (upstream of Crowleytown), Batsto, Wading, and Oswego rivers (Burlington County); Great Egg Harbor River above Mays Landing (Atlantic and Burlington counties; pH now regularly exceeds the 5.5 criterion, given watershed development activity over the past 20 years)

There are, surprisingly, thirty-two waterbodies or portions of waterbodies in New Jersey that are designated "wild trout streams."

Illus. 10. The Wading River at Route 563 in Jenkins' Neck—a quintessential "Pinelands Stream"

They are mostly little first-order feeder streams to rivers like the Musconetcong, Black, or Pequest, but one (Van Campens Brook) flows straight to the Delaware, and another is actually the uppermost reach of the mighty Passaic River. All are classified "trout production waters" (therefore only artificial lures or flies are permitted); none are stocked; all enjoy excellent water quality and substrates supportive of freshwater shrimp and clean-water insects; and none produce trophy-size fish. Most would be categorized as "freestone," some portions "meadow" brooks. Fishing these streams is an art, and short light tackle a must. They are streams for those not into an effortless guaranteed catch, and landowner permission should be sought on some stretches.

- Average Width: < 15 feet
- Average Depth: < 2 feet
- Current Velocity: moderate to high
- Temperature: cold to cool
- pH: ± 7
- Turbidity: low
- Substrate: rock, pebbles, gravel
- Pool/Riffle Ratio: ≤ 1:1
- Land Use Development: ranges from public forests to a few secondary roads
- Principal Target Species: brook trout
- Examples: Dark Moon Brook/Bear Creek (Sussex County); Flanders and Trout Brooks (Morris County); Passaic River headwaters (Morris County); Van Campens Brook (Warren County); West Brook (rainbow trout) and Pequannock River (Passaic County)

STANDING WATERS.

In New Jersey, classical definitions of standing (lentic) waterbodies are frequently confounded by the fact that most of these waterbodies, irrespective of size, were created by *impounding* them behind dams. Water flows into them, through them, and (usually) over the dam to the river below. Still, they are relatively easier to classify than flowing waterbodies. From a scientific standpoint, several schemes have been used, usually in conjunction with one another. First, they may

Illus. 11. Bear Creek at Bear Creek Road in Warren County—a meadows-type "wild trout stream"

be classified as to whether they are of natural or man-made origin, but that in and of itself is more significant to the geologist or "limnologist" (one who studies fresh waterbodies) than the average angler. Likewise, classification into categories based upon their seasonal mixing patterns and/or stage of eutrophication could be the subject of a whole book. Suffice to say that most of New Jersey's lakes and ponds fall under the category of being nutrient-enriched (eutrophic), and those deeper than about 10 feet experience mixing both before and after, but not during, summer ("dimictic").

Without getting into the physics, what happens in many lakes with the onset of summer is that surface waters warm rapidly and don't mix effectively with deeper waters, setting up what is called a *thermocline*— a layer of water through which temperatures drop very rapidly (more than about 2°F/yard). Above the thermocline remains a warm zone called the *epilimnion*, below it the *hypolimnion*, which can't exchange gases or nutrients with the surface despite wind action. This stratifi-

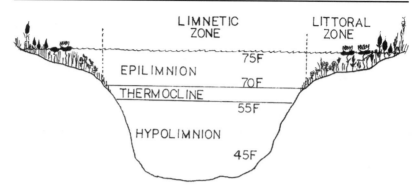

Fig. 3.20. Summertime lake stratification and zonation.

cation can be important to fish and fisherman alike, by determining where some species may be found and whether they can survive a hot summer. For example, though trout require a cold-water zone during summer, it does them little good if they don't have sufficient oxygen, which is what can happen if the hypolimnion is below the depth that light can penetrate to stimulate oxygen production by plants. At such times you will be more apt to find a trout *in* the thermocline (which may be more than just a few feet deep) or near the mouths of inflowing streams, the latter observation being made by Luftglass and Bern (1998) in regard to Spruce Run Reservoir. Other than that, thermal stratification is of little import to shore fishermen or boat anglers fishing the "littoral" (nearshore) zone for largemouth bass or pickerel. Boat anglers in search of cold- or cool-water species found in the "limnetic" (open-water) zone, however, can increase their chances if they employ an electronic fishfinder and temperature meter.

I have chosen a fairly pragmatic classification scheme based on the size and chief purpose of the waterbodies. I simply call them *lakes*, *reservoirs*, or *ponds*. Like Armantrout's (1998), my definition of a lake is a waterbody that is at least 20 acres in size; a pond is one smaller than 20 acres. My demarcation between lakes and reservoirs is based on whether they were initially conceived for drinking water supply purposes or (as in the case of Merrill Creek Reservoir) low-flow augmentation to the Delaware River. These reservoirs are all deep (maximum depths > 30 feet) and exceed 300 acres in surface area. I have rendered general descriptions and examples of representative water-bodies in each category, but three other publications are also worth getting. One is the *Fish and Game Digest* (which annually lists stocking schedules, quotas, size limits, and special regulations concerning "Trophy Trout" and "Holdover Trout" lakes); another is the DEP's (1998c) *Inventory of N.J. Lakes and Ponds*; and still another is Perrone's (1994b) book entitled *New Jersey Lake Survey Fishing Maps Guide*.

Of some *lakes* (waterbodies larger than 20 acres), only twenty-three exceed 100 acres, and most have maximum depths of less than 15 feet, making them warm- or maybe cool-water fisheries. The deep lakes, suitable for trout during summer if the epilimnion doesn't become depleted of oxygen, are in that area circumscribed by the Delaware River and Routes 78 and 287. The problem of oxygen depletion, mentioned earlier relative even to a reservoir as deep as Spruce Run, is common also to comparatively shallow Union Lake in Cumberland County, where a thermocline sets up at the depth of about 10 feet. Lakes with maximum depths of less than about 8 feet generally do not experience summer thermal stratification, and, unless advanced eutrophication and excessive blooms of "planktonic" (drifting) algae limit light penetration to the bottom, oxygen concentrations remain suitable for warm-water fish during summer.

Lakes have more open-water volume than ponds, where the littoral zone may encompass most of the waterbody's acreage. The food chain in the open-water (limnetic) zone of large, deep northern lakes is comprised of planktonic algae, microscopic invertebrates with names like "copepods" and "cladocerans" (a.k.a. water fleas), golden and some other species of shiners, trout, and walleye. Alewives have been planted in some to serve as additional food for sport fishes. Channel catfish are found deeper, and bass, muskies, and pike can be found near drop-offs and moving into littoral areas to feed at dawn and dusk. Most of our deep lakes are capable of supporting what is called a "two-story" fishery composed of cold- and warm-water species. Most fishing is done with trolled spoons, live herring or shiners, or jigging. Live-bait fishing and jigging can be done in summer, and through the ice in winter. In mid– and southern New Jersey lakes, the limnetic zone comprises a lower percentage of the surface acreage relative to that of the littoral zone, and the water is shallower. Large-mouth bass, pickerel, and tiger muskies are the primary predators of the limnetic zone in those lakes.

Starting at the shoreline, unless intentionally carved out and deepened, a healthy littoral zone begins with a ring of *emergent* plants like cattails and pickerelweed, proceeds through an area of rooted plants with floating leaves (lily pads), and terminates in a zone of sub-mergent plants including a variety of "pondweeds," watercress, water milfoil, and strands of attached algae. The plant community is very diverse here, including everything from little floating ones with root-lets dangling below their leaves ("duckweed") to a form of algae that only lives on the backs of snapping turtles. The invertebrate life is similarly diverse, including clams, snails, crayfish, insect larvae and adults like dragon- and damselflies (the latter being a more slender, dainty, and pretty form of the same taxonomic order), water striders (or "skaters" you see slipping across the water surface), and "water boatmen" and "whirligig" beetles. Two nice illustrated guides to these plants and animals are *Pond Life* by George Reid (1967) and Need-ham and Needham's (1962) *A Guide to the Study of Fresh-Water Biology* (both have likely been reissued since I acquired mine).

The littoral zone is a place that provides habitat, refuge, and food for a variety of minnows, which are eaten by sport fishes. What you fish with is a function of what sorts and sizes of species you want to catch. As a general rule you'll catch just about any nonbenthic species (yellow perch, largemouth bass, crappies, bluegills, pickerel) with a spinner; bass family members with a bobber and worm or a fly "pop-per"; large "bigmouth" bass with rubber worms and large plugs; pick-erel family members with large spoons and spinners or live shiners; and catfish or carp with baited hooks fished near bottom.

Of the hundred or so places listed in the Division of Fish, Game and Wildlife's guide to places to fish, I picked several that offer shore fish-ing, boat ramps, and (as of 1994) boat rentals, noting species stocked there in the nineties:

- Lake Mercer (Mercer County): 275 acres, stocked with tiger muskies and channel catfish.

- Budd Lake (Morris County): 376 acres, shallow throughout, stocked with tiger muskies, pike, and largemouth bass.
- Lake Hopatcong (Morris and Sussex counties): 2,685 acres, stocked with trout, muskies, tiger muskies, walleye, striped bass–white bass hybrids, and channel catfish. Since most of the lakeshore is private property, shore fishing is pretty much limited to the state park, but there are numerous mercantile boating facilities.
- Lake Musconetcong (Morris County): 329 acres, stocked with trout, but reported to have a "murderous population of chain pickerel" (NJDEP 1998c).
- Shenandoah Lake (Ocean County): 50 acres, stocked with trout and tiger muskies.
- Greenwood Lake (Passaic County): 1,920 acres, stocked with muskies, tiger muskies, walleye, and channel catfish.
- Pompton Lake (Passaic County): 204 acres, stocked with trout, pike, and channel cats.
- Shepherd Lake (Passaic County): 74 acres, stocked with trout.
- Cranberry Lake (Sussex County): 179 acres, stocked with trout.
- Swartswood Lake (Sussex County): 494 acres, stocked with trout, walleye, and channel catfish.
- Lake Waywayanda (Sussex County): 255 acres, stocked with trout.

There are many other lakes that provide one or more kinds of access, including cartop launching areas. These are also catalogued in the fish and game division's list of places to fish, along with fish types and quality of fishing there. I have selected examples of these lakes based upon either my own experience or facts provided in the literature which I felt justified singling them out:

- Lenape Lake (Atlantic County): This is a 350-acre impoundment of the Great Egg Harbor River at Mays Landing. Being of Pinelands drainage, it is cedar water. Its average depth is 6 feet, but a band of 8- to 12-foot depths runs through its linear course toward the dam. All opportunities but livery facilities are available, and it boasts of big bass and pickerel. Largemouth have been stocked.

- Scarlet Oak Pond (Bergen County): I picked this one because it is unusual for New Jersey, having originally been a mining quarry. As such, it is unique in terms of depth for such a small acreage (22 acres, which is why per my definition it had to be included under "lakes"). Its maximum depth is 54 feet; its average is 21 feet; and it can hold trout year-round. Shore fishing is the only type permitted, but it has a full complement of trout, largemouth bass, crappie, yellow perch, bluegills, channel cats, and bullheads.
- Union Lake (Cumberland County): At one time this was the largest man-made lake in the United States. It was constructed in the 1790s by damming the Maurice River at Millville, and, with 898 acres, it is South Jersey's largest lake. According to the state's inventory of lakes (NJDEP 1998c), the bulwark of

Illus. 12. Lake Hopatcong looking south toward the north end of Byran Bay

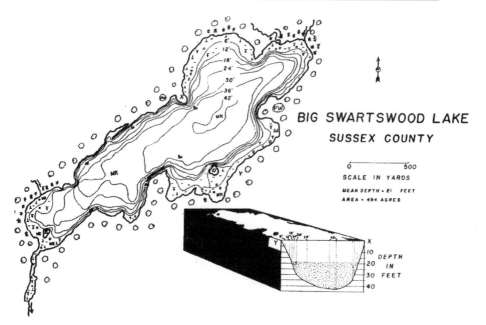

Fig. 3.21. The depth contours of Big Swartswood Lake in Sussex County. Courtesy NJDEP.

the sport fishery is largemouth bass, though pickerel, channel cats, and other species are also abundant. Access is good from shore, and there are both municipal and state boat ramps.

• Sunset Lake (Cumberland County): This is an 88-acre impoundment on Cohansey Creek. It provides all types of access but does not have a boat livery. The average depth is 4.7 feet, the maximum (one small hole near the dam) is 10 feet. Channel cats and tiger muskies have been stocked, and, according to Stan Haines's "Quick Tips" in Perrone's (1994b) book, Sunset Lake "is a terrific lake to take a kid fishing."

• Carnegie Lake (Mercer County): A long, ribbonlike impoundment of the Millstone River where it flows past Princeton University, this lake averages 4 feet in depth but has a 6- to 8-foot channel running through its length. The lake is 237 acres in size, is stocked with channel catfish and tiger muskies, and offers shore, cartop, and ramp access.

• Assunpink Lake (Monmouth County): This is a 225-acre im-

poundment completed in 1975. It is a "Trophy Bass Lake"
with average and maximum depths of 5 and 14 feet, respectively. Part of the Assunpink Wildlife Management Area, it
offers shoreline access and a boat ramp (oars or electric power
only, and anglers may find good pickerel and crappie fishing in
addition to largemouth bass fishing.

- Deal Lake (Monmouth County): A stone's throw from the
ocean, this wishbone-shaped lake of 158 acres is connected
to the Atlantic by a flume and herring run. It is stocked with
trout in spring, as well as channel cats, tiger muskies, and
pike. Being very eutrophic, its average depth is 5.4 feet, but
a couple of 9-foot holes exist. Shore, cartop, and ramp launching digs are available.
- Parvin Lake (Salem County): Another "Trophy Bass Lake"
offering shore and cartop access and (as of 1994) livery facilities. The 95-acre lake, in Parvin State Park, also offers good
catfish and sunny fishing.
- Surprise Lake (Union County): This is a 25-acre lake that I
used to fish regularly as a young adult, and it is still reported
by the NJDEP to have an excellent population of pickerel,
which readily snatch a Johnson Silver Minnow with pork rind
attached, not to mention bluegills and largemouth bass,
which love taking fly poppers. Surprise Lake is easy to fish
from shore, and rowboats were available at least as of 1994,
but the lake cultivates a pretty good growth of weeds during
summer. It is now being dredged to help matters.
- Furnace Lake (Warren County): This is a deep lake (average
depth is 18 feet; maximum of 37 feet) of 53 acres, created
originally for flood control. It is stocked with channel cats,
largemouth bass, tiger muskies, and trout, offering fishing for
all seasons. All but livery options are available.
- Mountain Lake (Warren County): I threw this in because it
is a *natural* lake of 122 acres with a maximum depth of close
to 40 feet. Its mean depth is 17 feet, and it offers year-round
fishing for brown and rainbow trout, plus largemouth bass,
crappies, and yellow perch. It has extensive littoral *and* lim-

netic habitat, and offers both boat ramp and shoreline fishing access.

I count eleven waterbodies in my *reservoir* category. Five are owned and operated by the state (Merrill Creek, Monksville, Manasquan, Spruce Run, and Round Valley reservoirs); four by the Newark Watershed Conservation and Development Corporation (NWCDC; Canistear, Clinton, Oak Ridge, and Echo Lake reservoirs); and one each by United Water (Oradell Reservoir) and the New Jersey District Water Supply Commission (Wanaque Reservoir). These are deep waterbodies. All except Echo Lake exceed 40 feet in depth, and two (Round Valley and Merrill Creek reservoirs) are well over 100 feet deep. All except Manasquan Reservoir, which is in southern Monmouth County and has a maximum depth of 40 feet, are in the northern part of the state, where they support a great trout fishery. Spruce Run and Manasquan reservoirs offer the most diverse fisheries. Merrill Creek and Round Valley reservoirs offer lake trout fishing and are, in fact, designated "Trophy Trout Lakes," while Spruce Run and Manasquan reservoirs have also been stocked with hybrid striped bass described in my next chapter. Spruce Run, Manasquan, Canistear, and Clinton reservoirs offer the most shoreline diversity and lit-

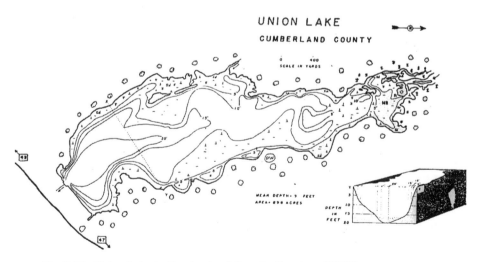

Fig. 3.22. Union Lake in Cumberland County. Courtesy NJDEP.

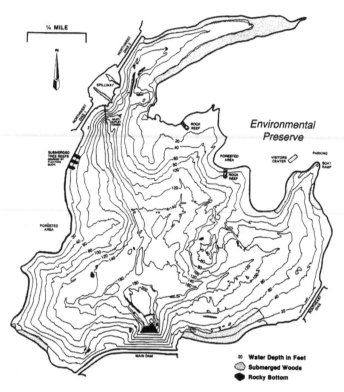

Fig. 3.23. Merrill Creek Reservoir, New Jersey's deepest inland waterbody. Courtesy NJDEP.

toral zone habitat, the rest providing a greater amount of limnetic zone habitat. You only need a New Jersey fishing license and trout stamp to fish the state's reservoirs, but you need a daily or seasonal permit (in addition to daily boat ramp reservations) for the NWCDC's reservoirs (obtainable at their Newark or Newfoundland offices), and permits from United Water or the New Jersey District Water Supply Commission to fish (shore only) the Oradell and Wanaque reservoirs, respectively. Luftglass and Bern (1998) provide good accounts

94

of fishing in the state-run reservoirs, and Perrone (1994b) and the NJDEP (1994) provide specs and/or fishing tips.

Here are condensed descriptions based on those publications and (in the case of Spruce Run, Round Valley, and Canistear) my own observations:

- Merrill Creek Reservoir (Warren County): 650 acres; maximum depth 210 feet (the deepest of our waterbodies); stocked with lake, brown, and rainbow trout, smallmouth bass, plus alewife for forage; young, deep lake with limited productivity; lake trout not yet realizing their growth potential because of such curbed productivity and forage, my son Geoff's catch being an obvious exception!
- Manasquan Reservoir (Monmouth County): 770 acres; 40-foot maximum depth; stocked with trout, small- and largemouth bass, striped bass hybrids, tiger muskies, and channel cats. Development plan included many habitat enhancement measures, including installation of gravel spawning areas and artificial brush shelters, and leaving in place 108 acres of timber and 28 acres of stumps. My 29-year-old son, Bruce, has fished for bass there, and he loves it.
- Monksville Reservoir (Passaic County): 505 acres; maximum depth 90 feet; stocked with trout, walleye, and channel catfish; benchmark reservoir for walleye.
- Spruce Run Reservoir (Hunterdon County): 1,290 acres; maximum depth 80 feet; constructed in 1963; stocked with brown and rainbow trout, striped bass hybrids, pike, tiger muskies, and largemouth bass; excellent shoreline fishing, wading, and camping; nice trout feeder streams on its northern side.
- Round Valley Reservoir (Hunterdon County): 2,350 acres (second only to Lake Hopatcong in surface area); maximum depth 160 feet; constructed in 1963 and opened in 1972; stocked with lake, brown, and rainbow trout. State record lake and brown trout, smallmouth bass, and eel were caught there, and I caught rock bass there from the pebbly shore of one of the relatively remote tenting areas.

- Canistear Reservoir (Sussex County): 350 acres; maximum depth 42 feet; stocked with trout, walleye, and smallmouth bass; midway up "Can-U-Steer" road from Route 23 to Highland Lakes.
- Clinton and Oak Ridge reservoirs (Passaic County): Each more than 400 acres and 40 feet in depth; both stocked with trout and smallmouth bass.
- Echo Lake (Passaic County): 300 acres; maximum depth 32 feet; stocked with smallmouth bass and trout, but also known for its largemouth bass, crappie, and yellow perch fishery.

Oradell and Wanaque reservoirs are 620 and 2,310 acres in size, respectively. They offer shore fishing by permit, probably yielding good catches of warm-water species (I haven't been there), but neither is stocked by the state.

Pond ecosystems resemble those of the littoral zone of lakes, and, although onetime plantings of trout are done in some of them in late March, and channel cats are stocked in certain of those larger than

Illus. 13. My son Geoff holding a beautiful lake trout taken from Merrill Creek Reservoir

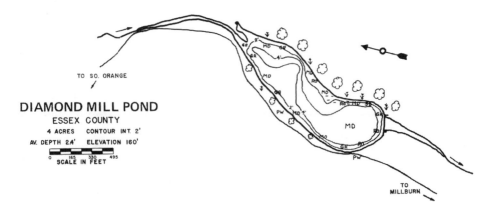

Fig. 3.24. The depth profile of Diamond Mill Pond in Millburn. Courtesy NJDEP.

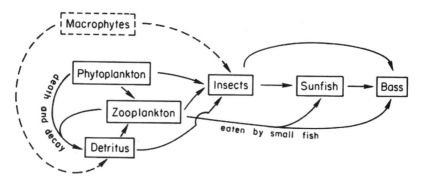

Fig. 3.25. A sunfish-bass food chain. From *Water Quality in Warmwater Fish Ponds* (1979), 67. Reprinted courtesy of Claude Boyd.

ten acres, pickerel and largemouth bass are usually the top predators and sport species. Often the limnetic zone is insignificant, if not almost nonexistent. The bluegill–largemouth bass pairing is a popular pond combination from Georgia to Massachusetts. Their food web is illustrated in figure 3.25. Diamond Mill Pond (ill. 14) is known for stocked trout fishing in spring and big carp.

Some ponds, possibly created initially to be centerpieces for planned-unit developments (PUDs) where lawn fertilization is the rule rather than the exception, may receive inordinate amounts of contaminated rainwater and storm sewer runoff. These ponds can be-

Illus. 14. Diamond Mill Pond in Union County on opening day of trout season, 1998

come saturated with nutrients and sediment, often losing their whole contingent of emergent aquatic plants and becoming dominated by submerged vascular plants and attached algae, requiring you to strip weeds off your hook after every cast. Nevertheless—in testimony to the pliability of largemouth bass, sunfish, crappies, yellow perch, and pickerel—some of these ponds can produce an abundance of fishes, many of them unexpectedly large. That's partly due to the fact that oxygen concentrations usually don't drop to intolerably low levels in summer, although winterkill can occur if the pond freezes, gets covered with snow, and oxygen is consumed by decomposer species more quickly than it can be replenished by plants.

Of 134 waterbodies less than 20 acres in size, I've listed 24 that offer cartop launching and shore fishing. Again, lots of others with public shore fishing access are included in the NJDEP's (1998c) list of places to fish, and you should be able to find bluegills, largemouth bass, and perhaps pickerel in most any of them (* denotes state ownership):

98

- Burlington County: Smithville, Woolman's, and Sylvan lakes (the last two stocked with largemouth bass in the nineties).
- Camden County: Cedar,* Hirsch's, Hopkins, and Oak* ponds (the last one stocked with channel catfish) and Evans Lake.
- Cape May County: Tuckahoe Lake.*
- Cumberland County: Bennetts Mill Pond,* Mary Elmer Lake (channel cats stocked).
- Gloucester County: Greenwich and Swedesboro lakes (both stocked with channel catfish).
- Hunterdon County: Amwell Lake* (stocked with channel cats).
- Middlesex County: Davidson's Mill Pond (stocked with channel cats).
- Ocean County: Lower Shannoc Pond.*
- Passaic County: Brushwood Pond.*
- Salem County: Riverview Beach Pond, Woodstown Memorial Lake (stocked with channel catfish).
- Somerset County: Powder Mill Pond.
- Sussex County: Lake Ocquittunk* and Sawmill Lake.*
- Warren County: Ghost Lake* (stocked with largemouth bass).

As you can see, though protection of freshwater fishing habitats and places to fish will be a never-ending battle in the face of competing interests, New Jerseyans still have many opportunities to enjoy a wide variety of species, some in the trophy-size range. Now let's take a look at all of our saltwater fishes and places where they can be caught!

CHAPTER 4

Saltwater Fisheries

I n 1996 New Jersey ranked third among the Atlantic and Gulf states, behind only Florida and Texas, in numbers of saltwater anglers (841,000), numbers of saltwater fishing trips (9,892,000), economic impact ($1.5 billion), and numbers of jobs generated by saltwater fishing (16,112) (Maharaj and Carpenter 1997, cited by Camhi 1998). 'Nuf said.

The Fishes

According to Able and Kaiser (1994), New Jersey's estuaries are either full- or part-time home to 98 species from 46 families. They list some of the same species that the NJDEP does on its freshwater checklist, in particular clupeids, so there's some double-counting between lists. Able and Kaiser's (1994) inventory doesn't include species typically found exclusively in the ocean, such as the Atlantic cod, ling, tunas, billfish, and most sharks, so the number of distinct sport fishes roaming our coastal waters must number more than 120 species. I've included discussion of 40 species representing 25 families. One species I chose to include here, rather than under freshwater fishes, is the eel, in reference to which I must introduce the term "catadromous." This is defined by Armantrout (1998) as having a "life history strategy, that includes migration between fresh- and saltwater, in which fish reproduce and spend their early life stages in saltwater, move into freshwater to rear as subadults, and return to saltwater to spawn as adults."

John Geiser, fishing editor for the *Asbury Park Press*, once wrote (ca. the seventies) a great little paperback about New Jersey's salt-

Illus. 15. Flags welcoming anglers and tourists alike at Liberty State Park in Jersey City

water fishes called *The Shore Catch*, which has been out of print for some time. I lost that book moving from Fair Haven to Highland Lakes, New Jersey, or maybe from Highland Lakes to San Francisco, but somehow I managed to retain his whole series of news strips upon which the book was based, those having been published in 1969 in the former *Asbury Park Evening Press* and mailed to me at UMass by Mom. I've made liberal reference to information in those stories, as well as many by Captain Al Ristori, who writes for the *Newark Star Ledger* and contributes to *The Fisherman*. Hopefully, my accounts and tales do them and the fish justice.

In writing these sections I drew upon data published and accessible via internet from the National Marine Fisheries Service (NMFS) commercial landings database and Marine Recreational Fishery Statistics Survey database, the latter of which was begun in 1979. Commercial landings are based on dockside records, while recreational estimates are based on phone surveys, party boat records, and some on-site "creel census." I have reported mostly what is called "catch" in the sections to follow, in order to provide some feeling for trends in abundance, however soft. Catch is what anglers claim they *caught,*

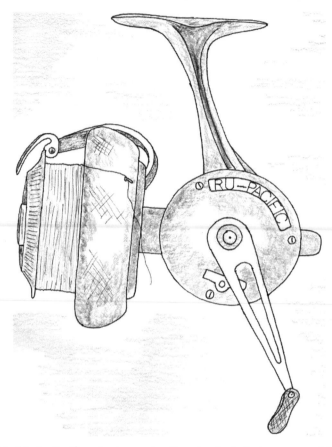

Fig. 4.1. My Ru-Pacific spinning reel, a French winch that was one of the first spin-casting reels to enter the spin fishing market here in the fifties.

not what they actually took home, which is called "harvest." Thus, catch should not, at least for some species, be interpreted and compared directly with commercial landings. Also, while catch and landings provide some gauge of population trends if levels of effort expended in pursuing each species remain constant, those numbers can't be relied upon if effort changes. That is why fisheries scientists rely more on ratios of catch *and* effort, as explained in chapter 5. And, last but not least, landings statistics may be capped by quotas and/or underreported by both commercial and sport fishermen.

A major difference between saltwater fishing and freshwater fishing is that the former is a domain frequently shared by sport and commercial fishermen from Florida to Massachusetts, and some in Canada, in addition to fishermen from overseas. Therefore, protection (let alone management) of these species is difficult at best, and usually fraught with controversy. Where specially applicable, I've dealt with that subject, but bear in mind that some pending management actions discussed here will, in one form or another, have been acted upon by 2000. Meanwhile, let me get on with the description of our "common wealth in ocean fisheries," to use a phrase from the title of a book by Christy and Scott (1965).

STRIPED BASS (*MORONE SAXATILIS*).

New Jersey's contribution to California. Back in 1897 and 1899, tanks filled with a combined total of 435 bass yearlings were shipped there via train and released into San Francisco Bay (Laycock 1966). I remember reading somewhere that they were snatched from the Navesink River near Red Bank, but in any event, they are thought (McLaren et al. 1981) to have been progeny of Hudson River stock. A wide-open niche was apparently available for the taking, because the striper population exploded on the West Coast. A web site (www.bayareafishing.com/stripers.htm) is even now devoted to striped bass fishing in the Bay area.

The striped bass can be found throughout New Jersey from the New York state border on the Hudson to Trenton on the Delaware. It can easily be identified by its shape and the seven to eight narrow longitudinal stripes along the upper part of its sides. It is an inshore fish. Although it reigns supreme in the surf, it also likes to congregate around habitat-forming structures such as jetties, bridges, and piers. Those of you who were here in the late seventies may recall that bass habitation among the old, dilapidated piers on the lower west side of Manhattan was one of the major issues that killed the proposed Westway. Striped bass are a favorite quarry of surf casters from North Carolina to Plum Island, Massachusetts. Down south, they are called "rockfish," and years ago up north they were familiarly known as *Roccus* (which had been their genus name until 1966, when taxonomists showed that, by the rules of nomenclature, they properly belonged under the genus *Morone*). Some fishermen's passion for the old name was so great that, even after the American Fisheries Society officially recognized the name change (AFS 1970), which many sport fishermen fought vigorously, the legacy of *Roccus* was immortalized on beach buggies and boats still bearing names like "Roccus Chaser."

The striped bass has two "cousins" that are popular among some sport fishermen in New Jersey. One is the white perch (*Morone americana,* described in the next section); the other a cross between the striped bass and the white bass (*Morone chrysops*). This hybrid, raised at the Hackettstown hatchery, has been stocked in Spruce Run and Manasquan reservoirs, Lake Hopatcong, and Assunpink and Cranberry lakes. Like the tiger muskie, described earlier, the bass hybrid was bred to capitalize on the size and fighting traits of one (striper) and the habitat type of the other (white bass). While the hybrid may get quite big (state saltwater record is 12 pounds 7 ounces; freshwater record is 10 pounds 14 ounces), its maximum size falls below that of striped bass. It can also be distinguished by breaches in the stripes below its lateral line.

Like the American shad, the striped bass is an anadromous species. It spawns on both sides of New Jersey, passing New York City on its way up the Hudson River and Philadelphia on the Delaware River. Spawning typically occurs during the first weeks of May in fresh water just above what is referred to as the "salt front." This is an area where fresh and salt waters mix along a wedge-shaped, deeper prong of salt water moving in and out of the estuary with the tides, while the lighter fresh water flows downstream atop the wedge. Naturally, the exact location of the salt front varies with the tide, but it is typically well downstream of places where the shad chooses to spawn.

After fertilization in the milieu of gametes scattered widely throughout the water, the eggs pass through a larval stage of two to three months, when they become juveniles (finding themselves farther down the estuary and much reduced in numbers). Mortality estimates vary, but a figure of 25 percent of their total numbers per day from fertilized egg to juvenile striper (McFadden 1977) puts the challenge they face in perspective. Survivors grow about 5 inches during their first year and may be a foot in length by age two. Legal-sized bass (28 inches in New Jersey) may be seven to eight years of age, the really big ones (25- to 30-pounders) being in excess of fifteen years. In general, striped bass in the 1- to 2-foot range are called "schoolies"

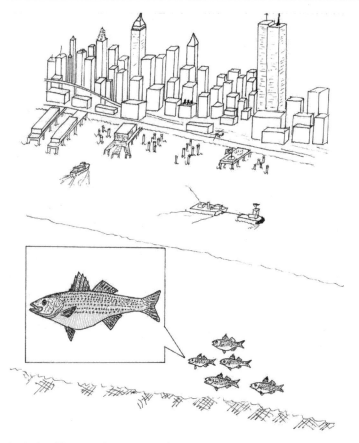

Fig. 4.2. Striped bass on their way up the Hudson to spawn above Kingston, N.Y.

(after their schooling behavior), while those in the 40-pound range are dubbed "bulls" (even though most are probably females, or "cows," the majority of which live longer and get bigger than the males). The state (and in fact world) record bass taken by hook and line in salt water was one 78-pound-8-ounce specimen caught near Atlantic City in 1982. The biggest one caught in our fresh waters (Delaware River, 1995) was an even 35 pounds.

Together with bass from the Chesapeake Bay system, the stripers we catch here from late spring through late fall are part of a communal assemblage of fish that, after spawning in their home rivers, enrich our coast with a wealth of sport fishing thrills. This virtue, their need for rivers with lots of freshwater flow and habitat to spawn in, and their table value (from a restaurateur and commercial fishing standpoint), also makes them a complex resource to manage. The economic importance of the striped bass to the commercial fishermen and their markets is undeniable, yet the sport fisheries also contribute substantial sums to state and local economies. The ASMFC estimated that in 1995 1 million anglers took 7 million trips and spent $160 million in pursuit of striped bass (*The Fisherman*, January 23, 1997).

Striped bass have historically been subject to large swings in abundance everywhere along the Atlantic coast. Fishing pressure notwithstanding, these fluctuations also have to do with the success (or lack thereof) of individual "year classes" of fish produced in the spawning grounds. Some year classes fare far better than others, often dominating the catch for many years as they grow and mature. In general, however, populations (as graded by landings) increased from the fifties through the mid-seventies; decreased to fifties levels in the late seventies; and crashed in the eighties, resulting in formulation of the first striped bass FMP by the ASMFC in 1981 and passage of the Atlantic Striped Bass Conservation Act in 1984. To illustrate, the sport catch was estimated to be 37½ million pounds in 1960, 56½ million in 1965, 73 million in 1970, and only 8½ million in 1979 (Malat 1993). Fortunately, as a result of harvest quotas and size limits on commercial and recreational fisheries dictated by states and the federal government (the latest FMP was put out in 1998), striped bass abundance has steadily increased. The recreational catch was restored to some 30 million fish in 1996, and concomitant increases were also seen in commercial landings in states that still allow those forms of bass fishing (New Jersey and Pennsylvania are not included in that category).

The years 1997 and 1998 saw an even greater abundance of stripers, bordering (in November 1998) on the implausible ("best day of bass fishing in 50 years," according to Al Ristori in his November 8, 1998, *Newark Star Ledger* account of the previous day's surf casting; "wall-to-wall," according to Bruce, who lives near the North Jersey shore). Stripers were caught regularly from spring through early winter from Palisades Interstate Park in Bergen County to Bayonne in Hudson County; throughout Raritan Bay from Hazlet to Sandy Hook;

Illus. 16. Spring striper fishing at the Chart House restaruant wharf on the Hudson River in Weehawken

in 15 to 40 feet of water just offshore around landmarks like the Outer-bridge Crossing and "Shrewsbury Rocks"; on beaches and jetties all along the surf; inside inlets and bays including Barnegat Inlet, Great Egg Harbor, the Mystic Island area of Great Bay, and Corson's Inlet in Strathmere (Cape May County); and in the mouth of the Maurice River. Striped bass were still affected by the earlier pollution block in the Delaware, but by the early nineties there were reports (Johns 1994b) of big stripers as far upstream as Phillipsburg in summer, contrary to conventional fisheries biology wisdom.

Most anglers I have watched along the Hudson seem to favor hooks baited with sea worms or "bunker" chunks, while Delaware River anglers report good catches on live sand eels as well. Results of the 1998 Long Beach Island surf tournament revealed that 52 percent of the stripers were caught on bait, but, in good weather and clear water, surf and tidal creek fishermen also like to toss metal lures, popping (surface) or swimming (deeper-running) plastic plugs that imitate baitfishes. (Before plastic, these plugs were made of wood, their effectiveness being measured by the scars in their paint. You may find these in some antique shops.)

Boat anglers may find that hooks baited with live eels, herrings, or mackerel work best for catching bull or cow bass. In fact, the largest striper I ever hooked into was one I enticed with a "live-lined" 12-inch mackerel fished from shore on the Piscataqua River in New Hampshire many decades ago (alas, I lost it to a lobster pot tether after an hour's battle). One thing you may observe, and have to respond to inventively, is the sometimes whimsical feeding behavior of the bass. Bigelow and Schroeder (1953) note that when stripers are feeding amorously on one kind of prey, they often ignore others. I found proof of this once in Great Bay, New Hampshire, while tracking a striper to which we had fastened a sonic tag. I chanced upon a huge group of big bass essentially straining the surface of the water for what is termed the *epitokous* stage of the sea worm *Nereis*. These fat, pinkish little reproductive contrivances were swarming all over the surface during their nuptial dance, and the bass were literally bouncing off our boat. Still, the fish would touch none of my varied offerings. Similar pickiness was reported in the news in November '98 in accounts of stripers focused on anchovy and "peanut bunker" in the surf. Despite all this, though, in most instances you'll catch your fair share of bass when they are nigh in.

Unfortunately, striped bass have a habit of accumulating PCBs

Fig. 4.3. *Epitokous* stage of the sea worm. After Barnes (1963, 201).

through the food chain in some of their natal waterbodies, and consumption advisories have been issued by state departments of public health. The first advisory, and concomitant ban on commercial fishing, was imposed by New York state a quarter century ago. Fortunately, with improved water quality, the New York State Department of Environmental Conservation announced on February 22, 1999, that levels of PCBs in striped bass in the lower Hudson River were finally below action levels and that the health department might consider lifting its ban. Good news, though that will no doubt lead to renewed conflicts over allocations between sport and commercial fishermen once again. As this book went to press, the ASMFC had already begun to push for further restrictions on recreational catches—from two to just a single striper per day in coastal waters.

Further information specifically about striped bass and how to catch them is available through the Fisherman Library, including Caputi's (1993) book *Fishing for Striped Bass*, Captain Al Lorenzetti's (1995) video *Live Bait Striped Bass*, and Frank Daignault's (1996) book *Striper Hot Spots: The 100 Top Surf Fishing Locations from New Jersey to Maine*.

WHITE PERCH (*MORONE AMERICANA*).

This is the fish for those who like light tackle fishing for pelagic species and can't wait for trout season to start. It is also one you can catch from shore in relatively sheltered areas, making it a good choice for kids.

The white perch is a smaller, though still scrappy, relative of the striped bass. Its genus name was also changed from *Roccus* in 1966, but probably few anglers noticed or cared. Although seasonal presence and abundance receives good coverage in *The Fisherman*, my records show that white perch are paid little if any attention in the standard news media. It is now and then referred to as a "sea perch," presumably to distinguish it from the yellow perch, even though they look entirely different. White perch are olive to silvery gray above, silvery white on the bottom, and have no stripes (neither vertical, like yellow perch, nor horizontal, like striped bass).

The ones you might expect to catch average less than 10 inches in length, but they get as big as 15 inches or so, weigh up to 3 pounds, and

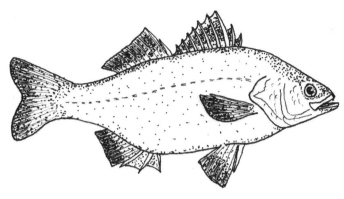

Fig. 4.4. White perch, a great little sport fish common to New Jersey's tidal creeks.

live for possibly more than ten years. The state record is bifurcated — the freshwater record is 3 pounds 1 ounce (Forest Hill Lake, 1989); the saltwater is 2 pounds 12 ounces (Little Beach Creek, 1998).

White perch are found in our intertidal creeks and bays from Cape May County to Hudson County on the Atlantic and Hudson River side of New Jersey; from the Maurice River to Trenton and occasionally to the mouth of the Musconetcong River (110 miles upstream; Beck 1995) on the Delaware River side; and in landlocked waterbodies as large as Lake Hopatcong (which in 1950 produced the [then] state record fish of 2 pounds 8 ounces) and as small as (or smaller than) Red Bank's Shadow Lake (which was connected to the sea by the Navesink River prior to the damming of Nut Swamp Brook).

In nonlandlocked situations, white perch is what is called a "semi-anadromous" species — semi because it makes spawning runs to fresh or slightly salty water in spring but doesn't go out to sea after spawning — rather retreating to deeper parts of bays in summer and fall and congregating again at the mouths of tidal creeks in winter prior to spawning again. In late fall the white perch may be caught in deeper holes of estuaries like the Mullica River. By mid-January they start showing up in nice numbers in locales in the Toms and Metedeconk rivers. In March they provide good sport from Weehawken on the Hudson to the South River off the Raritan, throughout all of the unobstructed tributaries of Barnegat Bay from Bricktown to Toms River, in the Mullica and Egg Harbor rivers, in the lagoons and back bays inside Surf and Sea Isle cities, and around the horn into all of the streams tributary to the Delaware estuary. The January 16, 1997, edition of *The Fisherman* singled out the Maurice River at Millville as being synonymous with white perch, noting that while spring and fall may provide the best action, you might catch them year-round.

The National Marine Fisheries Service has estimated that New Jersey's saltwater fishermen caught between 14,000 and 105,000 white perch a year from 1990 through 1996, averaging about 34,000 a year when you discount the two extremes. Adult white perch (5 to 6 inches long) feed on worms, shrimp, killies, and spearing, the best baits being grass shrimp, pieces of worms (earthworms in fresh water, bloodworms in salty water), or spinners and streamers that imitate baitfishes.

BLUEFISH (*POMATOMUS SALTATRIX*).

The bluefish is a most popular sport fish, partly because of its prevalence and abundance, and partly because of its voraciousness and probability of being caught on any given day during summer and fall. Bigelow and Schroeder (1953) called it the "most ferocious and bloodthirsty fish in the sea," and Ristori (1995) referred to it as an "animated chopping machine." In fact, adult bluefish are commonly called "choppers" (their young are labeled "snappers"). The bluefish is the only member of its family (Pomatomidae). Its closest relatives belong to the family Carangidae (jacks and pompanos, not generally common in New Jersey and therefore not covered herein).

Bluefish are found along the North Atlantic coast from Massachusetts to Florida, in the Gulf of Mexico, along most of the South Atlantic coast of South America, all around the African continent, and encircling Australia (Ristori 1995). In New Jersey it has been dubbed the number one species in terms of recreational catches, although this varies from year to year (ranging from almost 17 percent of our marine sport catch in 1990 to less than 7 percent in 1996). On a coastwide basis, New Jersey ranked number one in recreational catches of bluefish from 1987 through 1996 (22 percent of the estimated total catch), and, when asked, most anglers between here and Cape Cod have apparently identified bluefish as their primary target (Ristori 1995). My own creel survey experience on Cape Cod and Buzzards Bay in the late sixties (Piehler 1972) and news accounts of angler preferences during simultaneous blitzes of striped bass and blues in 1998, however, argue that most people claim to be fishing for striped bass.

Like the striped bass, bluefish populations have experienced their ups and downs in abundance. Along our coast they were very abundant in the mid-1600s, vanished by the mid-1700s, rebounded in the mid-1800s, all but disappeared during the first half of the twentieth century, bounced back again from the late fifties through the eighties, declined in the early nineties, and resurged again in 1997 and 1998, to the point where they had become, according to many popular accounts, an annoyance to striped bass anglers. The sport catch constitutes some 75 to 80 percent of the combined recreational and commercial landings in the mid-Atlantic, partly because bluefish do not freeze well (Ristori 1995). As an example, in 1986 (a peak year for bluefish) the New Jersey recreational harvest was about 131 million pounds, while the commercial fishery recorded only 3 million pounds (constituting less than 3 percent of the total commercial fishery landings of all species in New Jersey).

The bluefish has a torpedo-shaped body that is flattened a bit on its sides. Its dorsal half is bluish gray, while its bottom half is more silvery. The lower jaw juts out from the upper, and both are equipped with razor-sharp teeth. It has small, almost imperceptible scales. It lives an average of fifteen years, but may survive for twenty and grow to 3½ feet in length and well over 20 pounds in weight. The state record is a 27-pound-1-ounce fish caught in 1997 on "Five Fathom Bank." Of interest is the fact that until 1972, when a fish of almost 32 pounds was caught in North Carolina (Bulloch 1986), the previous world record (a 27-pounder taken off Nantucket, Mass.) had stood since 1903! Such a bluefish may be twenty years of age, whereas a 1-footer ($\approx$ 2 pounds) is about two years old, a 2-footer ($\approx$ 6 to 8 pounds) about five years old.

The bluefish is a migratory species. Tracking waters with temperatures higher than about 55° F, it is found south of Virginia during winter, sometimes well south (Bigelow and Schroeder [1953] reported the recapture of a bluefish tagged in New York in 1936 off the coast of Cuba in 1939). The snapper blues found in our bays and estuaries during summer are by and large the offspring of adults that

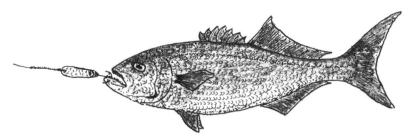

Fig. 4.5. Bluefish latched onto a Hopkins 4.

spawned in April and May along the inner edge of the Gulf Stream off the Carolinas. They enter our bays in late spring and early summer when they are about 2 inches long and, feeding on small shrimp and resident baitfish, grow to 7 or 8 inches by the time summer is over and they begin heading south. In mid-summer, schools of snappers averaging 5 inches or so have addicted many a youngster to the joys of fishing at almost no capital cost to the parents. I still think that fishing for snappers with an inexpensive bamboo pole, bobber, and hook baited with a "spearing" (or silverside, described later) is more fun than tackling them with more sophisticated gear.

If you chance upon very small snappers late in the season, it is likely that you have taken one of the progeny of spawning activity that occurs along our continental shelf during summer. Long thought to represent a separate and distinct spawning season, resulting in recruitment of two sets of juveniles into the population, contemporary studies suggest that spawning may occur in sequential pulses from spring through early fall and that the apparent hiatus in production of juveniles could be an anomaly caused by differential survival of larvae (New Jersey Sea Grant Pub. No. NJSC-94-286). Nonetheless, most of these summer descendants evidently remain offshore and begin moving south in fall when they are 2 to 3 inches in length (Bulloch 1986).

The bluefish is a fish that young anglers can grow up with, widening their chase of choppers through habitats requiring more skill and equipment such as casting in the surf, trolling and jigging just outside the breakers, or tackling the "jumbos" in deeper water. These exploits get more expensive. Nonetheless, no matter how many Hopkins lures I lost to the bluefish's teeth or an intemperate whip of my surf wand

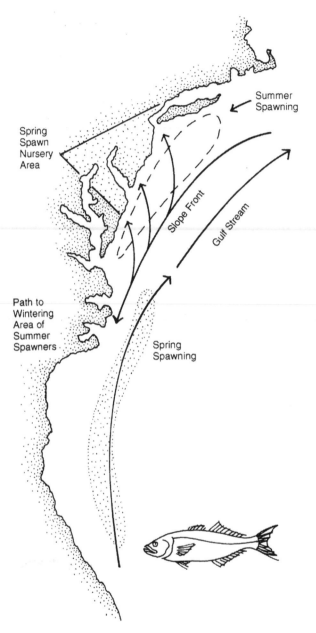

Fig. 4.6. Bluefish spawning areas off New Jersey. From Ristori (1995), *Fishing for Bluefish.* Courtesy of the Fisherman Library Corp., Point Pleasant, N.J.

and Ru-Pacific winch (an early and still top-notch spinning reel that was made in France, and fortunately is still in my possession), I never tired of these animal fights. Fortuitously, the bluefish were so voracious that even an old tablespoon handle fitted with a treble hook served the purpose when Dad could afford no more Hopkins that week!

At this stage in life the bluefish is a very opportunistic eater, pursuing squid, menhaden, butterfish, mullet, bay anchovy, river herrings, and anything else of the right size. Usually, the presence of bluefish ripe for the taking is first signaled by an alertness to seagulls flocking over and diving into the sea. They are pursuing the same prey from above as the bluefish are from below, and it is not uncommon to see butterfish, mullet, or other quarry being tossed onshore by the surf as these fish desperately seek refuge. This is when Doug and I would interrupt our body surfing to retrieve our rods from their sentinel sand spikes and start tossing something, it didn't matter what, into the fray. If you can cast even 30 yards into the surf you are bound to hook one with every cast; whether you land one depends on

Illus. 17. Doug (*left*) and the author with a stringer of bluefish caught in the Ocean Beach surf, ca. 1954

whether you've included a steel leader between your lure and the end of your line. Bluefish schools may feed intermittently all day long, although the best time to catch them is at dawn and dusk.

You can catch bluefish from the bulkheads of Weehawken, Hoboken, and Jersey City along the Hudson River, all the way around Cape May Point, throughout Great Bay and Barnegat Bay, and into the saltier portions of rivers tributary to these bays (like the Manasquan River) or the ocean. A sampling of fishing reports from *The Fisherman* bears witness: October 1995—bluefish up to 15 pounds being taken at the Chart House restaurant (on a long wharf in Weehawken) and Liberty State Park (Jersey City); June 1996—bluefish heavy in the Arthur Kill and lower Hudson, in the Manasquan inlet (especially around dark), "super" around Barnegat Light, and "great nightly action" at Cape May; and November 1998—"hammering monsters" just outside the (Barnegat) inlet, and "Cape May rips on fire!" Thus, when you hear that the "blues are running" (as will be reported in the Sunday editions of any major New Jersey/New York newspaper), you generally won't have to travel too far to find them. To learn more about blue-fishing techniques read Captain Al Ristori's (1995) book *Fishing for Bluefish* or view Lorenzetti's (1996) video *Fishing for Bluefish*.

MACKERELS AND TUNAS.

This family, Scombridae, is officially recognized as "mackerels" (AFS 1991). I tacked on the tunas because I wasn't sure how many readers might be aware that they are members of the same family. Anyway, I picked two to illustrate—bluefin tuna (*Thunnus thynnus*) and Atlantic (a.k.a. "Boston") mackerel (*Scomber scombrus*)—and three others to mention and characterize—little tunny (or "false albacore," *Euthynnus alletteratus*), bonito (*Sarda sarda*), and Spanish mackerel (*Scomberomorus maculatus*). Of these species, I have caught representatives of only two, and I am not sure which of two possibilities the second one was because I lost it. I was casting a Hopkins 4 toward a school of jumping, tunalike fish in the Ocean Beach surf with my first rod and reel, which was sans star drag. I think what hit my lure must have been a little tunny because, as Geiser (1969) notes in his species account, its initial 50- to 100-yard run "melted my spool"

116

(which had less than 100 yards on it, and my reel handle was spinning so fast that I could not break the fish's run before witnessing the last inch of line peel off). The next day Dad bought Doug and me each a Penn Beachmaster, which did have a star drag. Equipped with the new reel, though, I never hooked into another little tunny, which can get as big as 25 pounds (approximate state record, caught off Sea Bright in 1977). Little tunny and bonito still frequent our coast, an abundance of the former taking mullet at Sandy Hook in late September 1998, according to the *Newark Star Ledger*; some 18,000 of the two were estimated to have been caught by New Jersey anglers in 1996.

The bluefin is the biggest, by far, of the lot. My story about star drags notwithstanding, Geiser's (1969) account of bluefin mentions that, until the star drag was invented, few bluefin were caught by sport fishermen. This is supported by an article in *The Fisherman* (January 14, 1999), which attributes the lack of a record heavier than 705 pounds in 1933 to the inadequacy of the "thumb-stall" reels of the times. The New Jersey record is a 1,030-pounder caught off Point Pleasant in 1981, while historic chronicles show that 1,500- to 1,800-pound bluefin have been caught by commercial means. Smaller bluefin can be differentiated from bonito and little tunny by the fact that their steely bodies have no markings, and their bellies and entire trunk are fully scaled. Little tunny have naked bellies, wiggly stripes on top with spots beneath, whereas bonito have curved, longitudinal stripes on their dorsal half. In addition to other features and markings, all of these species can be distinguished from Atlantic mackerel by the fact that there is no space between their two dorsal fins, and that they have eight, not five, of those little "finlets" you see between their dorsal and tail fins.

Bluefin, also known as "horse mackerel," have been described as "oceanic wanderers" by Bigelow and Schroeder (1953). As they travel the seas, they prey upon smaller schooling fishes such as herrings, mackerel (their own family members!), and silver hake (whiting). They have no serious natural enemies, other, perhaps, than large sharks. School bluefin in the 15- to 300-pound range reportedly overwinter in the Sargasso Sea, a productive region of the Atlantic due east of Palm Beach and just north of the Tropic of Cancer, but generally do not spawn; giants (>300 pounds), however, overwinter in the

117

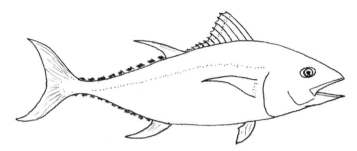

Fig. 4.7. The mighty bluefin tuna.

Caribbean and spawn in March and April in the Gulf of Mexico (Bulloch 1986). They start working their way north past Florida in mid-May, then New Jersey, and, by June, showing up around Cape Cod and Massachusetts Bay. During their northward migration, small to medium-large bluefin travel in small schools, frequently jumping synchronously. Bluefin can be found throughout summer and fall off the New Jersey coast, but you will need to charter a sport fishing boat suitably outfitted for the task if you don't own one yourself. Although Bruce reported losing one in the hundred-pound range in 60 feet of water off Seaside Park in November '98, most big fish are taken at or beyond the 20-fathom line. Daily and weekly reports about charter boat action and target species can be accessed on the internet by using either a general search (e.g., "New Jersey sport fishing reports") or going to something more specific such as (in 1998) "NauticalNet. com" and then clicking on "charter boat reports."

Many years ago the bluefin, because of its size and habit of ruining nets set for smaller fish that bluefin prey upon, was considered a nuisance to commercial fishermen in New England (Bigelow and Schroeder 1953). That changed in the fifties and sixties with the rising demand for tuna steaks, canned tuna (though albacore is now used), sushi dishes, and other cuts of tuna tartar, and stocks of other fish beginning to dwindle.

The account of the bluefin tuna sport fishery given by Barrett in his January 14, 1999, issue of *The Fisherman* caused me to go back and edit this chapter after I thought I had completed it. He provided documentation of a tremendous New Jersey–based bluefin charter boat fishery in the thirties, so colossal that Brielle went on record to call itself the "Fishing Capital of the World." In one summer week in 1939, over 11,000 bluefin were caught in New Jersey, one-tenth of that catch being landed in Brielle. Then, over the next three decades, with commercial purse seining added to traditional long-lining methods of exploiting bluefin, stocks suffered severe depletion in the early seventies. Commercial and recreational limits were then instituted for bluefin harvests, and sport fishermen who venture sufficiently offshore that they are not confused with bluefish anglers are now required to obtain a tuna fishing permit from the NMFS's Washington, D.C., or Gloucester, Massachusetts, offices. International mandates are established by the International Commission for the Conservation of Atlantic Tunas (ICCAT), which ruled in 1981 that harvest of bluefin should be reduced to as near zero as feasible in order to allow the population to rebuild.

Both sport and commercial catches are regulated by the "Highly Migratory Species" (HMS) FMP mandated by the Magnuson Fishery Conservation and Management Act. As with many of these plans, sport and commercial interests disagree on quotas, especially since the annual quota for sport fishermen is now about what was landed in that one week in 1939. The HMS FMP is further complicated by what is called "bycatch" of nontarget species, in particular, sharks. The subject is taken up later, but, in brief, bycatch is synonymous with *dead discards* (Crowder and Murawski 1998).

Not being a big game fisherman myself, I would refer you to Barrett's (1992) book *Fishing for Tuna and Marlin* or Metcalf's (1994) video *Fishing for Giant Bluefin Tuna* for more detailed information on catching bluefin. I am more intimately acquainted with the Atlantic mackerel. Actually, both Atlantic and Spanish mackerel invade our coast from time to time during spring and fall, 1998 being one of those years. The Spanish mackerel is the bigger of the two, and New Jersey holds (again) both the state and world records for that species (9 pounds 12 ounces, caught off Cape May in 1990). The state record for Atlantic mackerel is 4 pounds 1 ounce (Manasquan Ridge, 1983). King mackerel (*Scomberomorus cavalla*), also found here sometimes, is what the *New York Daily News* edition of September 25, 1998, must have been talking about when it reported 19-pounders being taken in the Shrewsbury River (the state record is 29 pounds, off Beach

119

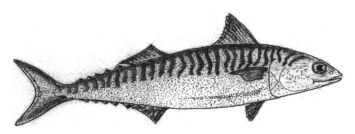

Fig. 4.8. Atlantic (a.k.a. "Boston") mackerel.

Haven in 1987). Spanish and king mackerel actually look more like bonito, except that they have spots, not stripes.

The Atlantic mackerel is a pretty fish, with its vermicular (worm-like) green-on-silver dorsal markings and torpedo shape. It is not a fish that you can just decide to go and catch whenever you want to, though. As Bigelow and Schroeder (1953) put it—"mackerel are where and when you find them." I have had better luck finding them in the bays and harbors of Massachusetts and New Hampshire during summer than at any other time in our waters, my experience being that the "Bostons" tend not to enter our tidal waters as much when they are here (May and again December). They are accessible, how-ever, by boat, and the best strategy is to keep your nose in the news-papers, eyes on the water, and a diamond jig in your tackle box.

Atlantic mackerel are also wanderers. They overwinter in the depths off the Virginia Capes (Geiser 1969) and spend the rest of the spring and summer spawning over the shelf in waters of at least 46°F (which is in May in New Jersey) as they make their annual trek north. The young feed on practically any floating animals except jellyfish and have sometimes been found to have gorged themselves on an in-vertebrate scientifically named *Calanus* (a "copepod") but popularly known to commercial fishermen as "red feed" or "cayenne" (Bigelow and Schroeder 1953). By the end of their first fall season they may be almost 10 inches in length ("tinker macks"), by their eighth year al-most 17 inches.

Scomber scombrus populations are also characterized by wide fluctu-ations in abundance. Bigelow and Schroeder (1953) reported swings in Massachusetts landings from 100 million pounds in 1885 to an av-erage of only 37 million pounds a year from 1933 to 1946, and our own recreational catches have varied from 5 million pounds in 1981 to just over 2,000 pounds in 1988 and to just under a million pounds in 1997. New Jersey's commercial landings since 1980 were lowest in 1980 and 1985 (1.6 and 1.8 million pounds) and highest in 1991 and 1996 (more than 18 million pounds). It seems clear that the commercial fisheries are much more adept at catching schools of these fish than is the recreational fishery (which, as I said before, is generally hit or miss).

Everybody knows what tuna tastes like in one or more of its culinary forms. The mackerel has been described as being "OK" if cooked properly, but also oily and requiring expeditious refrigeration to prevent spoiling, which can cause sickness (Bulloch 1986).

SUMMER (*PARALICHTHYS DENTATUS*) AND WINTER (*PSEUDOPLEURONECTES AMERICANUS*) FLOUNDERS.

"Flatfishes" and their relatives are all flat, oval, with both eyes perched on the topside of their adult bodies. They are brown on top (looking down on them as they skulk on the bottom of the waterbody) and white on the underside. The shade of brown that they adopt is a function of the color and intensity of the type of bottom they are lying upon or hidden in with their eyes peeping out. The side they lie on depends upon the family into which they have been born.

The species described here — one a "lefteye" (or "left-handed"), the other a "righteye" (or "right-handed") species — represent two of the three families of flatfishes recognized by the American Fisheries Society. Bothidae is the family that includes the fluke; Pleuronectidae the family that includes the winter flounder, yellowtail flounder, and Atlantic halibut (the biggest flatfish, growing to upwards of

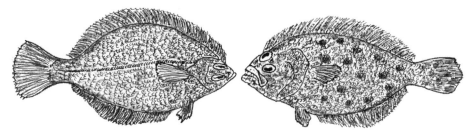

Fig. 4.9. New Jersey's principal flatfishes: winter flounder (*left*) and fluke.

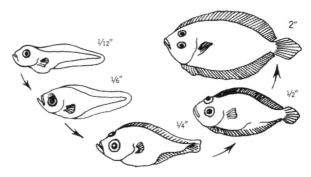

Fig. 4.10. Larval development of a fluke showing migration of its right eye over the top of its head to the left side of its face. From Alice Jane Lippson and Robert L. Lippson, *Life in the Chesapeake Bay*, 2d ed., 123. Copyright © 1997 by Alice Jane Lippson and the Johns Hopkins University Press.

500 pounds); and Soleidae the family that includes the hogchoker (*Trinectes maculatus*), which is listed here not because it is a sport fish, but because it is a curious smaller relative that is common to estuaries on both sides of New Jersey. Each member of these families shares a common bond. When they are larvae, one of their eyes migrates to the other side of the head so that both eyes are on the same side of, say, one cheek. Coloration follows. The side that they lie on turns white; the side facing up turns brown (providing a selective advantage, from an evolutionary standpoint, given that their profiles don't stand out as well to predators peering down from above).

The summer flounder, or fluke, as it is commonly known, is the left-eye species. During early development, the eye that would have been on the right side of a typical fish's head migrates over the top to the left side of its face. The fluke has numerous small, dark spots on its body, distinguishing it from a close relative, the "four-spot" flounder. Fluke are called summer flounder because that is when they appear in our nearshore waters. Our fluke spawn from late September to December during their offshore migration to overwintering grounds 25 to 80 fathoms (480 feet) deep. Spawning occurs at 60- to 150-foot depths (i.e., beyond the 10-fathom mark, which is about 5 miles offshore of Manasquan but some 15 miles out once you reach Cape May) and temperatures ranging from 54° to 66°F (Bulloch 1986). Fluke eggs and early larval stages are buoyant at first, living a pelagic lifestyle. As they develop into juveniles, they adopt a benthic habit, feeding mostly near the bottom but not averse to attacking prey (or artificial lures) well above.

Fluke begin showing up in our sport fishery during May—first the small ones (1 or 2 pounds), then the larger ones, which, at about 10 pounds, are also called "doormats" (for obvious reasons). The biggest doormats generally stay further offshore at 8 to 10 fathoms where they are fair game for commercial otter trawl fishermen. Honors for biggest fluke caught with sporting gear in New Jersey, a record held since 1953, go to a 19-pound-12-ounce specimen caught off Cape May, but nice-sized fluke may be taken in the surf and major inlets throughout late spring and summer. In June 1997 *The Fisherman* reported that they "were stacked up from Manasquan Inlet to the Route 70 bridge." Still, the best fluking can usually be had by boarding one of New Jersey's many party boats and making a day (or half day) of it, not having to worry about equipment, bait, getting "skunked" (no fish), or fileting your own fish.

It is no secret that, due to the taste and texture of their flesh, fluke have always been one of the most important sport and commercial species throughout their range, which for the most part extends from Cape Cod to South Carolina. So much so that in 1989 the fishery "crashed." From combined recreational and commercial landings of 15 million pounds in both 1983 and 1984, total landings dropped to 3.7 million pounds in 1989. Although recreational landings alone rebounded to 15 million pounds in 1998, and New Jersey's commercial fishermen were confined to landing 1.9 million pounds in 1998, the NMFS view that fluke are heavily overexploited and in need of more stringent bag, size, and indeed seasonal controls continues to be a source of strife among champions of sport fishing.

The winter flounder, sometimes called blackback, lemon sole, or mud dab, is a righteye species (the left eye migrates around the head during larval development). Winter flounder also have smaller mouths and do not grow as large as fluke. The state record is one 5-pound-11-ounce fish (a "snowshoe" flounder) taken off Barnegat Light in 1992, but they more commonly average about half a pound,

Illus. 18. The summer party boat fleet at Ken's Landing in Point Pleasant Beach

ranging from only 12 to 15 inches or 1 ½ to 2 pounds (Bulloch 1986). The winter flounder recently encountered a temporary name change. Historically named *Pseudopleuronectes*, it had the *Pseudo* dropped in 1984, but restored by 2000.

Winter flounder are uncommon south of New Jersey (Malat 1993) but provide great sport fishing and eating for New Jersey and New York anglers. They, as their name connotes, are found in our waters from late November through early May in a cold year, spawning in relatively shallow inshore waters starting in mid-December. Except for January and February, when the season is closed to protect spawning adults, they support a popular sport fishery from either shore or small boat in places like the upper Barnegat Bay and the Manasquan, Shark, Shrewsbury, Navesink, and Hackensack rivers. Come May, winter flounder start moving offshore, no doubt crossing paths with incoming fluke. Winter flounder are harvested by commercial fisheries in offshore waters during summer, and over the past decade New

124

Jersey anglers have caught about a million of the fish a year. This species is rated by the NMFS as being overfished but having the reproductive capacity to rebuild in ten years, given a recent increase in the minimum size limit and commercial net mesh size.

My first fluke fell for a piece of squid fished about 100 feet into the ocean at Ocean Beach Unit #1; my first winter flounder took a piece of bloodworm in a then undeveloped tidal channel off Barnegat Bay in Chadwick while Dad was painting and dewinterizing our cottage in April. The best ways to catch flatfish involve baited rigs, but winter flounder (having smaller mouths) require proportionately more petite hooks and smaller pieces of bait than fluke. Fluke can be had with bloodworms, clams, killies, silversides, or larger strips of squid or sea robin. The fluke will also usually strike the bait more vigorously, while the winter flounder is notorious for its light hits (*The Fisherman*, November 19, 1998, and my own experience). Kamienski's (1993) book *Fishing for Fluke*, his (1987) video *Fluke Fishing*, and Lorenzetti's (1995) video *Fluke Fishing—Improving Your Catch* provide popular accounts of that endeavor.

Finally, if you are wondering what happened to the hogchoker, it is a righteye species that is smaller than the others and distinguished by its rounded nose. It is apparently tasty but small and bony (Lippson and Lippson 1997) and is often sold in pet shops as "freshwater flounders" (Quinn 1997).

THE CODFISHES (GADIDAE FAMILY).

Except for Chicago in February 1982 (where I contracted frostbite for the first time despite being a lifelong fisher and skier), ground-fishing from a Brielle party boat in January was the coldest experience of my life. By the time we got to our fishing destination, our rods, reels, and line, which had been secured to the rails by pieces of string while we had coffee in the cabin, were encrusted in inch-thick cakes of frozen spray. An hour after the horn sounded, signaling that the anchor had been dropped and we could sink our lines, my hands were like claws, yet the fishing was good and a good time was had by all.

I count five members of this family among sport fishes I have known. They are the Atlantic cod (*Gadus morhua*), silver hake (*Mer-*

luccius bilinearis), red hake (*Urophysis chuss*), pollock (*Pollachius virens*), and tomcod (*Microgadus tomcod*). Let's consider the smallest one first this time.

The tomcod, also known as "tommycod" and "frostfish," is no longer subject to so much commercial pressure and is, therefore, not counted separately by the NMFS. Popular yet localized recreational fisheries exist where they are found in New Jersey estuaries, principally the Hudson and Hackensack Rivers (Able and Kaiser 1994). However, their numbers are insufficient to warrant inclusion in the NMFS recreational database, and their size and size variability are too small for the state to sanction a bulk record. A 9-inch tommy is considered a nice catch, but a 13-inch morsel is reported to have been caught at the old Chart House pier in Weehawken in 1997 (*The Fisherman*, March 27, 1997). Anglers I have spoken to in Hoboken love to eat tommycod, although some folks (e.g., Geiser 1969) feel that they are inferior to cod in taste.

The tomcod is an inshore fish, normally available to anglers only during its spawning season from November through March. Peak spawning occurs during late December and early January in moderately fresh zones of estuaries. Tomcod obviously look more like the Atlantic cod than any of the others pictured, but their markings, greenish color, and relatively small eye separate them from their larger relative, not to mention the fact that you are unlikely to catch a cod off the Hudson County wharves. Being bottom feeders, tomcod eat mostly small crustaceans, mollusks, and worms, but they provide good sport on light tackle.

Tommycod populations have been the subject of studies of the Hudson River due to concern about the effects of power plants, specifically *entrainment* of eggs and larvae into the pipes carrying river water that cools and condenses (for reuse) the steam powering the electric generating turbines, and impingement of juvenile and adult fish on the screens protecting those pipes from damaging debris. These concerns began in the late sixties with anxiety over potential losses of striped bass. The result was a negotiated settlement between the utility companies, the Environmental Protection Agency (EPA), and the New York State Department of Environmental Conservation that led to cooling-water intake design modifications and oper-

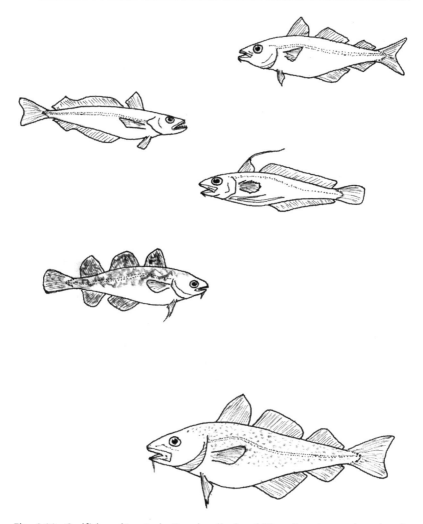

Fig. 4.11. Codfishes: (*top to bottom*) pollock, whiting, ling, tomcod, and cod.

ational changes to minimize impact on fish (AFS 1988). Whether related or not, given so many natural and man-made variables affecting abundance of our estuarine fishes, the winter '94/95 estimate of the Hudson River tomcod population (1.4 to 4.5 million) was three times that of '93/94 but still three times less than that of '82/83 (LMS 1995).

The rest of this group of species are found almost exclusively offshore in New Jersey, even though the pollock is frequently found schooling and feeding in a pelagic mode in the harbors of Massachusetts, where they are called "Boston bluefish." Geiser's (1969) remark on this geographical contrast in behavior, which is probably related to the combined differences in habitat type, water temperature, and distribution and abundance of prey species, confirms my own experience. I have only taken pollock on cod party boats in New Jersey, whereas, in summer 1979, my boys and I had a serendipitous, marvelous time hooking and releasing hundreds of pollock engaged in a feeding frenzy on river herring exiting a fish ladder (the "Herring Run") on Cape Cod Canal, Massachusetts. Our Hopkins 2s must have looked just like the small clupeids, and the Boston blues behaved exactly like their half-namesake of our waters!

The pollock has a greenish hue above and is silvery gray below its lateral line. It also has more of a crescent-shaped tail than its relatives. Pollock are found throughout the North Atlantic, even as far as the British Isles, where they are referred to as "coalfish." New Jersey is about as far south as they venture on this side of "the pond." They eat mostly small fish and pelagic shrimp, and are caught mainly during winter in New Jersey on the same grounds where cod and hake are found (i.e., > 50- to 200-foot depths). Clam bait and jigs work best, and don't be surprised if you hook into a big one. The New Jersey record remains one 46-pound-7-ounce fish caught on a party boat out of Brielle in 1975. That fish was considerably older than the 3-foot ones typically caught here (which are about ten to twelve years of age). Pollock spawn from late autumn through early winter, and are about a foot in length by age two, and 2 feet by age five (Bigelow and Schroeder 1953).

In New Jersey, recreational harvests of pollock far exceed commercial landings. Contrast an estimated 50,000 pounds caught by sport fishermen in 1987 to only a hundred pounds of commercial landings. The bad news is that in 1997 anglers are estimated to have taken only 625 fish (about 2,000 pounds), and commercial fisheries landed only 35 pounds (worth a total of $23 in the market). Both statistics reflect lower abundance of pollock in New England and the fact that New Jersey is at the lower extreme of the pollock's range (pollock tend to show up here when they are doing better up north).

The red hake, or ling as it is more commonly nicknamed, is distinguished from the cod, tomcod, and pollock by its two distinct dorsal fins (the second and third being connected) and from the silver hake by the long, filamentous (threadlike) extension of the third ray on its first dorsal fin and its long, modified pelvic fins, which form a bifurcated affair looking more like a musical "tuning fork" than a normal fin. The accepted (rather than colloquial) name for the ling used to be "squirrel hake," and its swim bladders historically were employed in the manufacture of what was called "isinglass" (a gelatinous substance formerly incorporated in clarifying agents and adhesives, according to *Webster's* dictionary). A unique habit of the ling is its penchant for hiding in the mantles (inside of shells, for all intents and purposes) of large scallops, enjoying what is called a commensalistic (mutually beneficial) relationship.

The ling is an exclusively American species. Its numbers too have dwindled over the past several decades, from commercial landings of 1.7 million pounds in 1980 to just under 150,000 pounds in 1996, and sport catches of 1 million fish in 1990 to only 60,000 in 1996. The state record is an 8-pound-12-ounce mouthful taken off Belmar in 1990. Ling and whiting could both be caught in spring and fall from the Long Branch pier before it was demolished, but, of course, they are still fair game for the winter party boat fleet. They spoil quickly, but taste good if quickly iced and prepared (Geiser 1969; Bulloch 1986).

Silver hake are more commonly called "whiting" and have been dubbed "as sweet a fish as one could ask for" (either fresh or smoked) by Bigelow and Schroeder (1953). Two to five million pounds a year were caught in New Jersey's commercial pound net fishery from 1942 to 1947 (fixed nets anchored just offshore that the fish unwittingly enter as they follow a lead rampart toward the pound), but just about two million pounds (fetching a modest $617,491) were landed in 1996 (now by trawlers, or "draggers"), according to NMFS statistics. The average adult is about 12 to 16 inches (1 to 1½ pounds) (Geiser 1969), and the state record is currently vacant (minimum weight of 2½ pounds being required to capture anyone's interest).

Whiting have no chin barbels, but they have lots of small sharp teeth. They prey on herrings and other schooling fish; Bigelow and Schroeder (1953) assert that a 23¼-inch whiting caught off Orient Point on Long Island had seventy-five 3-inch herrings in its stomach.

They may appear along the north Jersey surf in late fall after spawning offshore during summer, but most commonly fall victim to the sport angler's enticements dangled from a wintry party boat junket.

Last but not least, the biggest and, for most, the best eating of the family—the cod. Although the state hook-and-line record of 81 pounds has withstood the test of time since 1967, due to a century of overharvesting, a 37-pounder (also taken out of Brielle) was caught in March 1997, and the relatively newly fashioned wreck and reef habitats continue to provide predictable fishing for cods ranging from 5 to 20 pounds.

Cod spawn near surface during winter, an average female producing more than a million eggs, which, after fertilization, drift about in the ocean's currents for more than a month before eking out a benthic lifestyle once they are able to swim a bit. Their floating larvae feed on planktonic copepods, while juveniles and adults feed mostly on mollusks. Indeed, clams and squid are the most reliable baits for catching cod.

Other than large sharks and dogfish, man is the cod's worst enemy. As with most of our offshore groundfishes, the ranks of cod have been seriously depleted by commercial overfishing. Both big and small cod ("scrod") have supplied a rapturous market since colonial times, when they were elevated to pious symbolism. More recently, due to the decline in abundance and badly needed mesh-size restrictions and landing quotas, commercial catches in New Jersey have dropped from almost 80,000 pounds in 1989 to only 228 pounds in 1997 (worth a paltry $260 on the market). Similarly, the recreational catch was estimated to be only 950 fish for 1995, declining from nearly 29,000 in 1993. Hopefully, the NMFS management strategies aimed at rebuilding cod populations by the year 2002 will bear fruit, though bycatch of small fish caught by commercial flounder draggers remains problematic.

DRUMS (SCIAENIDAE FAMILY).

If you put your ear to a freshly caught constituent of this family, with one exception, you may hear a "drumming" or, in the case of croaker, a "croaking" sound. This is created by the quivering of muscles along the air bladder that causes it to rumble. Like cods, drums come in many shapes, sizes, and lifestyles. The biggest one is the black drum (*Pogonias cromis*), the state record being a 105-pound brute caught in Delaware Bay in 1995. The smallest is the spot (*Leiostomus xanthurus*). For some reason, unlike the case of the tomcod, the state keeps size records for the spot (8 ounces, Mullica River 1991). The most sporting is the weakfish (*Cynoscion regalis*), which also gets pretty big (the state record is 18 pounds 8 ounces, Delaware Bay, 1986). Finally, sort of intermediate in looks and habit—but subordinate in size (record being a 2-pound-3-ounce fish taken in Barnegat Bay in 1993) and lacking an air bladder with which to croak—is the northern kingfish (*Menticirrhus saxatilis*). The red drum (also known as "channel bass," ranging to 55 pounds), the Atlantic croaker (once very abundant in estuaries as far north as the Manasquan but now largely restricted to Delaware Bay), and the spotted sea trout (similar to the weakfish but not quite attaining their full-grown size) deserve mention in passing since they are worthy but less frequently caught species.

Black drum are generally caught by accident in the surf, but they are targeted by some anglers in Delaware Bay. The 105-pound record drum, which measured 54 inches, was caught near a place called "Slaughter's Beach" toward the Delaware side of the bay. It was apprehended by a drum-fishing apostle who is reported to have performed a "drum dance" on the stern of his boat to attract the fish before dropping his surf clam–baited hook (*The Fisherman*, November 19, 1998). That must be a good place to fish, because Cape May party boats make nightly drum-fishing trips into Delaware Bay during summer, and an 81-pounder was also caught there in June 1996. Seventy-three- and 68-pounders were caught during September in the surf at Mantaloking and Cape May, respectively (*Newark Star Ledger*, September 20 and 27, 1998), the former unexpectedly hitting a Bomber plug. Most, however, are taken with clams, generally in deep holes or sloughs in channels or inside sandbars.

Black drum have a humpback appearance; are dark gray with coppery red hues; and possess large, blunt, "pharyngeal" teeth in their throats (i.e., the pharynx) designed for crushing clams. They also sport a bunch of chin barbels for sensing bottom prey (their genus name *Pogonias* translates as "bearded"), and they make sufficient noise to have earned the species name *cromis*, which means "to grunt"

Fig. 4.12. Drums: (*top to bottom*) weakfish, spot, kingfish, and drum.

(Lippson and Lippson 1997). Geiser (1969) reports that youngsters (about a foot long) found during June in Delaware Bay and the mouth of Hereford Inlet (North Wildwood) had conspicuous brown and white bars on their sides. Black drum mature quickly (two to three years), when they are just above a foot in length, and they are very fertile, a single large female producing many millions of eggs over a life span lasting well over a third of a century. Only 996 black drum were caught by anglers in 1991, the only year for which an NMFS estimate was at hand, but commercial landings and young-of-the-year indices of abundance, derived annually by the Delaware Division of Fish and Wildlife, peaked, respectively, at 8,000 pounds and 0.8/tow in 1993 (up from 0.1/tow in 1992) (Michels 1995). I have never caught a drum, let alone eaten one, but the state record holder apparently shish kebabbed his, and Geiser (1969) notes that drum roe are "said to be excellent." Most anglers tend to release big drum unless they believe that their catch might challenge the size record.

Moving from the least abundant to generally the third most abundant of this family, we visit the little spot. Spot, sometimes called Lafayette, are most plentiful in Maryland and North Carolina, but they are also common in the Delaware estuary and (in a good year) may be found as far north and inland as Fair Haven on the Navesink River. They spawn in the ocean during winter, their young-of-the-year finding nursery habitat in estuaries from late spring through fall, with adults following to feed and fatten throughout the same period. They are short lived (maximum five years, more regularly two or three), and their abundance varies quite a bit in our area partly because we are at the northern, and environmentally more variable, fringe of their range. The years 1991 and 1994 were relatively good ones for spot, as gauged by New Jersey's recreational take of an estimated 108,000 and 304,000 fish, respectively, while 1993 was a very bad one (7,800). There is no directed commercial fishery for spot in New Jersey or Delaware, but spot commonly fall victim to shrimp and industrial fish trawls, and are then classified simply as bycatch. Most of this bycatch occurs in the southern part of their range, where some states are introducing requirements to reduce such losses with "bycatch reduction devices" (BRDs), which focus the gear more accurately on target species. Spot can be identified by their sloping heads, oblique dorsal stripes, and, of course, the spot behind their gill. The best fishing is at high tide, using shrimp or pieces of bloodworm, and they are tasty to boot. Mia seasoned, soaked in milk, coated with a mixture of egg dip and Parmesan cheese, and sautéed four in a pan for us last year, and they were superb.

The northern kingfish, another light tackle fish that can be caught during early summer and fall and is good to eat, is distinguished by its towering first dorsal fin, the dark vertical bars on its sides, and the chin barbel of a bottom feeder. I have caught kingfish in the surf and from nearshore party boats, but they also enter tidal rivers. *The Fisherman* (June 13, 1996) reported good catches in the Shark River near the Belmar railroad bridge as well as on the Atlantic City rock piles and in the inlets and surf near Sea Isle City and Wildwood. Although difficult to lose once hooked, because of their hard mouthparts, Geiser (1969) calls the kingfish a notorious "bait stealer"—quick, hard rap, kind of like how I would describe the chub's picking a salmon egg off a hook earmarked for trout. The NMFS estimates of recreational kingfish catches averaged about 200,000 fish from 1990 through 1996.

Finally, the sportiest of the group, the weakfish (a.k.a. "squeteague"). In the early fifties, when northern Barnegat Bay was undeveloped, eelgrass teeming with grass shrimp abounded, Brick Town was a mere gleam in developers' eyes, and even Toms River was a sleepy hamlet, my Uncle "Bud" would take Doug and me out in his 16-foot skiff (the *NanMar*, named after my cousins Nancy and Mary) to fish for weakies just inside Barnegat Inlet behind the lower end of Island Beach State Park. We started the day by hitching his lapstrake wooden skiff (fiberglass boats were just beginning to emerge on the scene) to a shrimp boat anchored inside the bay and buying a couple of pailfuls of shrimp to skewer onto long-shanked hooks, which hung about three feet beneath a bobber. Using freshwater spinning equipment, my first with a Mitchill Cap spinning reel that I wish I still had, we learned why these fish are also called sea trout and weakfish — the former being related to their looks and militant style; the latter to their flimsy mouthparts, which often surrendered the hook after a short fight.

Weakfish are found from Massachusetts Bay to southern Florida, although their presence in Mass Bay (as of 1953) was pretty much limited to one sudden and abundant materialization about the turn of the twentieth century (Bigelow and Schroeder 1953). Cycles also typify weakfish abundance; some (Geiser 1969; Bigelow and Shroeder 1953) have noted that common fisherman "folk wisdom" correlated these cycles to opposite patterns in bluefish abundance. If that were strictly true, we would not have seen the simultaneous bonanzas of 1997 and 1998 (*New York Post*, October 2, 1998—"weaks and blues blitzing everywhere"). In 1998 weakfish were taken all summer and fall from Montauk (Long Island) to South Jersey in the surf, Jamaica Bay, New York Harbor, Raritan Bay, the Manasquan Inlet, the Mantaloking and Forked river mouths on Barnegat Bay, Great Bay, the Mullica River, Corson's Inlet, and Delaware Bay. Bruce caught more than a hundred "weaks" in the Navesink River and the "rip" off Sandy Hook in '98. And, with improving water quality, dumping restrictions, and habitat protection, even Newark Bay and the Hackensack River are witnessing a comeback of weakfish, a 7-pounder reportedly having been caught near Berry's Creek (Lyndhurst) in 1994 (Quinn 1997). The AFMFC weakfish management plan for the Delaware estuary embraced maintenance and protection of essential spawning, nursery, feeding, and migratory habitat; confinement of dredging to seasons when weakfish are absent; enforcement of effluent limitations on wastewater discharges; regulation of oil and gas activities; and coordinated planning with neighboring states (Seagraves 1995). New Jersey, in 1998, imposed a 14-inch/14-fish limit on recreational catches, as well as size limits, quantity limits (150 pounds a day), and season closings on commercial gill net, pound net, and trawl fisheries.

Weakfish are a very handsome species with their iridescent purple, yellow, and greenish blue markings superimposed on a greenish gray background. They spawn near estuaries from May till July, depending on latitude, and their larvae and juveniles utilize those estuaries as nursery habitat during their first summer (Bulloch 1986). Juveniles feed on shrimps, worms, and other bottom-dwelling invertebrates as well as fish eggs and larvae (including those of their own species), and after six months are about 4 to 6 inches long. They mature between the ages of one and two and compete with bluefish and striped bass for baitfish like butterfish, herring, and sand eels, among others. By age five they may be about 21 inches long, weighing up to 2½ pounds, and when they get to be 3 feet long (10 to 11 pounds) they are dubbed "tide runners." Male weakfish make signature guttural sounds; females do not (Bigelow and Schroeder 1953).

BLACK SEA BASS, TAUTOG, AND SCUP— A TRIO OF "REEF FISH."

Three different families—but a popular threesome—of what Steimle (1995) described as part of a "guild" of structure-oriented fishes. From late fall through spring you'll see tautog (more commonly called blackfish) and sea bass listed on the same marquees as

Illus. 19. A winter blackfish and sea bass party boat anchored at Highlands next to the Route 36 bridge to Sea Bright and Sandy Hook

cods by party boat operators from Leonardo to at least Barnegat Light. Sea bass are also common joint targets of fluke hunters during summer and early fall. They are mainstays of New Jersey's in- and offshore reef fishery, begun in 1984 by the NJDEP's Division of Fish, Game and Wildlife Port Republic field station, which by the end of 1997 counted 1,117 tire, ship, and army vehicle reefs among its credits.

Both sea bass and tautog (frequently abbreviated "tog") are unique in appearance as well as biology—the former because they all start out in life as females, most changing to the male gender by age five (Bulloch 1986); the latter because of its propensity for plucking blue mussels, barnacles, and other normally atypical elements of a menu off rocks and other hard substrates (Himchak 1998). Both have benefited from creation of artificial reefs. Recreational sea bass catches doubled (to half a million fish) in the first half of the nineties, although New Jersey's commercial fishery has oscillated between 700,000 and 1,380,000 pounds a year from 1987 to 1997. Steimle (1995) notes that most sea bass are now landed as bycatch of the summer fluke fishery, and that the biology of this critter poses interesting management concerns. In particular, high fishing mortality of larger fish (mostly males) can affect the sex ratio and possibly even lower the age of hermaphroditic changeover, both with serious reproductive consequences.

Sport catches of tautog jumped from 161,000 pounds (about 134,000 fish) in 1981, two years before the reef program began, to over 2 million pounds in 1986, remaining between a half and one and a half million pounds since then. The newly arisen popularity and associated increase in angling effort directed toward tog, however, is not sufficient cause to blithely equate increased catches with an order-of-magnitude expansion of the population. In fact, in April 1998 the NJDEP announced an increase in the size limit from 13 to 14 inches (essentially protecting another age group of spawners) and quotas of ten fish a day from fall through spring and one fish a day in summer (when they spawn inshore).

Illus. 20. A ship being sunk to create an artificial reef off the Jersey coast. Courtesy NJDEP 1998e

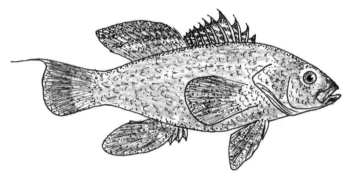

Fig. 4.13. Sea bass.

The black sea bass, *Centropristis striata*, belongs to the family Serranidae. Sea bass mature when they are 7½ inches, at ages two to three, and have a life span of fifteen years (Steimle 1995). In 1998 the NJDEP announced an increase in the minimum size for this species also, from 9 to 10 inches, and a short closed season (August 1–15). Sea bass average about 1½ pounds (slightly over a foot in length) but get as big as 8 pounds 2 ounces (the current state record, shared by a fish caught in 1992 at an inshore wreck and another taken in 1995 off Point Pleasant). Fish bigger than 5 pounds are generally referred to as "giants" in the popular press's accounts of weekly fishing action. Sea bass, togs, and porgies all produced great action from Long Island Sound through Jersey's reefs in fall 1998. As reported in three of our major metropolitan newspapers, sea bass and tog action remained good throughout the winter after one of the trio went deeper offshore and possibly farther south for the winter.

As mentioned before, sea bass are often caught in summer with the same rigs you would use for fluke. On a boat out of Highlands I chartered for a Taiwanese client years ago, my sea bass took the "pool" for the biggest fish (about 2 pounds) and my companion, who had never held a rod and reel before, took "high hook" honors with about a dozen fluke! We used the common fare of clams and squid for bait, but sea bass have also been known to chomp on a diamond jig occasionally (Bulloch 1986). Geiser (1969) remarked that you don't need much of a knack to hook sea bass, but you would be wise not to let them run, given their proclivity to dive into crevices.

Sea bass, being closely associated with the sea bottom, vary in color. Those I have seen are usually brownish, and their continuous spiny and soft dorsal fin, coupled with that threadlike extension of the top of the tail fin, serve to distinguish them from other species. They spawn in 50 to 135 feet of water in the ocean during spring and early summer, and both juveniles and adults are found inshore by mid-May, juveniles seeking an estuarine habitat in which to develop (Bulloch 1986). Migration to deeper waters, where they are taken on offshore wrecks and reefs, occurs in late October or early November.

The tautog, *Tautoga onitis*, is a member of the Labridae, or wrasse family. It is a big, long-lived fish—the state record being 25 pounds (off Ocean City in 1998). Togs love to be around reefs and rocky outcroppings, and delight in eating green crabs donated by sport anglers all winter long. This fish gets at least as much publicity in the newspapers during winter as stripers and blues during summer and fall. The October 12 edition of *The Fisherman* devoted a whole article to "Fall Toggin." In it, they singled out two areas in particular—the "17 fathom" area 8 miles southeast of the tip of Sandy Hook, and the "Mud Hole" 4 miles southeast of that. Togs are also taken in spring and late fall in Long Island Sound and around "the City Island Bridges" (*New York Post*, April 25 and November 20, 1998), as well as the jetties and inlets of New Jersey.

The state record for a tog was one caught off Ocean City in 1998. That fish was probably close to thirty years of age, because tog grow slowly. A one-year-old is only 3 inches; a six-year-old only 14 inches;

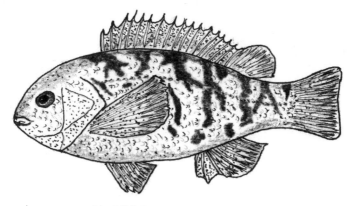

Fig. 4.14. Tautog, or "blackfish."

a ten-year-old only 19 inches; and a twenty-year-old only 22 inches (Himchak 1998). Tautog move inshore during spring to spawn and offshore in fall when the water temperature drops below 52°F. Other than for those inshore and offshore movements, they are not highly migratory. They are structure dependent throughout their lives, as exemplified by the fact that even their fry take up residence in empty shells or under rocks (Bulloch 1986) and their juveniles hang out in bays around pilings and the like (Himchak 1998).

Blackfish are tackle busters, mandating that you use heavy gear and sharp, sturdy hooks. Their firm white (delicious) meat makes for perfect baking and also fine chowder (Geiser 1969).

The scup (*Stenotomus chrysops*), better known to anglers as a "porgy," belongs to the Sparidae family. It was enormously abundant in Massachusetts in the late 1800s, but over a fifteen-year period suffered a cataclysmic decline (from 3½ million pounds to 200,000 pounds; Bigelow and Schroeder 1953). In the early sixties, Soviet trawlers were reported to have contributed to a reduction in our state's catch from 14 to 10 million pounds a year (Geiser 1969), and landings have continued to decline from some 5 million pounds a year in the early eighties to just over a million pounds in 1997. Estimates of recreational catches of scup have varied, almost cyclically, reaching 1 million fish in 1991 only to decline to a quarter million two years later, then rebounding to 1.6 million fish a year later — only to decline again to 144,000 in 1996. The Mid-Atlantic Fisheries Management Council's Summer Flounder Monitoring Committee (which covers sea bass and scup as well) reported a good '97 year class, but they increased the scup size limit from 7 to 8 inches (*The Fisherman*, November 26, 1998). Scup mature at two years of age ($\approx$ 6 inches; Bulloch 1986) and have the capacity to live fifteen or twenty years (Steimle 1995). Most range from half a pound to 3 pounds, but the state record is a 5-pound-14-ounce animal taken in Delaware Bay way back in 1976.

Scup spawn from May to August in inshore waters and the eastern end of Long Island Sound and Delaware Bay (Bulloch 1986). Juveniles seek shallow areas of bays and estuaries, but in summer even adults hug the shoreline so closely that a line drawn 5 to 6 miles out would enclose most of them (Bigelow and Schroeder 1953). Ac-

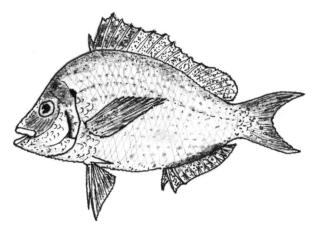

Fig. 4.15. Porgy, or "scup."

cording to Geiser (1969), they don't feed during spawning season but are easy to catch thereafter, hitting on small pieces of almost any bait available in the sea and providing a "whale of a battle." Scup are preyed upon by the usual big, pelagic, and aggressive species described earlier, and survivors migrate to offshore overwintering grounds in late fall. They also make delicious eating.

For further reading, Venturo (1995) has a book out named *How to Fish Wrecks, Lumps and Rock Piles,* and the Fisherman Press (1993) offers a compilation of their own stories entitled *The Guide to Blackfish and Sea Bass.*

SHARKS AND RAYS: THE CARTILAGINOUS ELASMOBRANCHIOMORPHI.

In contrast to all of the other species I have described, which belong to the taxonomic class Osteichthyes (bony fishes), sharks and rays are distinguished by the fact that their skeletons are composed of cartilage, which is lighter than bone. That was once thought to have been a primitive trait from an evolutionary standpoint, suggesting that other forms of fish evolved *from them*, but not so, according to the experts. Lagler, Bardach, and Miller (1962) write in their textbook *Ichthyology* that development of a cartilaginous skeleton is thought to have derived independently from, presumably, a common ancestor of all jawed fishes ("Gnathostomata"), a view reinforced

141

Fig. 4.16. Fossilized sharks' teeth found in a small creek in Manalapan.

more recently by Long (1995). Formerly classed as Chondrichthyes, meaning "cartilage fishes," they and their relatives the chimaeras ("ratfish") have since been reclassified as Elasmobranchiomorphi (meaning "plate gills").

Sharks were the first of the "modern" jawed fishes to evolve, most of which are the same as they were when they originated over 360 million years ago during the late Devonian period of the Paleozoic era (Long 1995). That period also witnessed the peak of a worldwide rise in sea level, which lasted into the Tertiary period of the Cenozoic era, 5 to 65 million years ago, which is why you can find fossilized shark teeth anywhere from creek beds in Monmouth County, New Jersey, to the black shales of Ohio and Pennsylvania. Rays date to the Jurassic period, which was sandwiched between the Tertiary and the late Devonian, 65 to 205 million years ago. Intrigued by an article that appeared in the *Asbury Park Press* in 1980 or thereabouts, I took my boys out to either Yellow or Manalapan Brook in Monmouth County (I don't remember which anymore), where, wading in knee-deep water with a kitchen sieve, we found a great many shark teeth in the gravelly sediment. I don't think they were planted there, but it surprised me to find so many that close to the surface!

Teeth are another distinguishing feature of sharks. They are a derivative of what are called placoid scales, or "dermal denticles," which protect the animal's skin and give it its characteristic roughness. Dermal denticles are also the roots of the shark's teeth, which are regularly replaced throughout life at a rate of about once a week in young fish and maybe up to 20,000 teeth over a ten-year period (Lane and Comac 1992). The spines of spiny dogfish and the "stingers" of stingrays are also derived from placoid scales (Long 1995). Sharks have no air (or "swim") bladders with which to facilitate buoyancy, but they do have a very large, oil-filled liver, which serves a similar function (Long 1995). Sharks and rays have either four or five pairs of gill slits, through which oxygen-laden water courses instead of being pumped through a single covered opening as in the bony fishes. They also have a series of unique *ampullae of Lorenzini* around their heads, which provide them with a sense of smell. Sharks are reported to be able to detect blood a mile away; vision takes over only when they are about 50 feet from their quarry (Lane and Comac 1992).

Sharks and rays further differ from any of the other fish included in this book in the fact that fertilization of their eggs occurs *inside* the females. Sharks are born live following a period of *in utero* development in an enlarged uterinelike oviduct, some getting their nourishment through a placentalike association with the uterine wall (*viviparous*) and others just milling about in an aqueous bath until the time comes to leave (*ovoviviparous*). Ovoviviparity is the rule (Lagler, Bardach, and Miller 1962). Skates and a handful of related sharks do it differently, being what are called *oviparous*. For them, development occurs in a purse-shaped egg case that measures about $1\frac{1}{2}$ x $2\frac{1}{2}$ inches not counting the horns. Anyone who has spent any time down the shore, especially after a storm, will recognize these brown cases that are also known as "mermaid purses" (Lippson and Lippson 1997). In all instances, the reproductive stratagem of the Elasmobranchiomorphi is the same: maximize fertilization efficiency, produce fewer young per litter, and protect them until they can fend for themselves. Not a bad plot of nature, but, combined with slow growth and maturity, it can lead to problems if they are harvested prematurely by man, their only enemy.

The spiny dogfish (*Squalus acanthias*) is well known to anyone who has ever taken a biology or comparative anatomy course. It is also the species you are most likely to catch unless you are specifically going after the bigger pelagic species like blue, mako, tiger, or great white sharks using big-game gear. Spiny dogfish should not be confused with the smooth dogfish of a different family. Spiny ones have spines at the front of each of their two dorsal fins and no anal fin. They are gray rather than brownish. In contrast to the spiny dogfish, the smooth

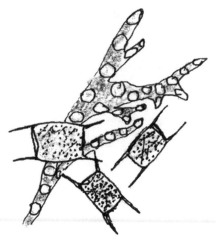

Fig. 4.17. Skate egg cases, or "mermaid purses," found on the beach after a storm.

dogfish has teeth that are flat and blunt. (Lippson and Lippson 1997). Spiny dogfish are ovoviviparous, and smooth dogfish are viviparous (Daiber 1995), which you wouldn't know just by looking at them.

Spiny dogfish are 8⅔ to 13 inches long at birth after being carried by their mother for 18 to 22 months (Bigelow and Schroeder 1953). They take at least ten years to mature, then produce litters of two to eleven "pups" about every three years (Camhi 1998). According to Bigelow and Schroeder (1953), they travel in packs and wreak havoc in schools of mackerel and similar fish, while also stopping to graze on crabs, squid, and worms as well. Spiny dogfish are summer visitors along the New Jersey coast but enter Delaware Bay in late fall and early spring when temperatures are below 60°F. Prior to 1990, dogfish and other sharks were harvested mainly by recreational fishermen. The biggest spiny dogfish landed in New Jersey by rod and reel (15 pounds 12 ounces) was caught off Cape May in 1990, and estimates of the total recreational catch range from 281,000 to 496,000 fish a year for 1990 through 1996. In 1990 the dogfish captured the attention of commercial fishermen and landings soared from 700 pounds the year before to 4½ million pounds. Spiny dogfish are caught mostly by gill netters and trawlers in New Jersey, which ranks fourth among all Atlantic and Gulf Coast states in landings of dogfish (and in fact all cartilaginous fishes). Their meat is mostly exported to Europe, their fins to Asia (Camhi 1998).

The sand tiger (*Odontaspis taurus*) was called simply a "sand shark" and listed in its own family of Charchariidae until the Commission on Zoological Nomenclature changed all that in 1965 (AFS 1970). In my catch, it typically ran second in abundance to the dogfish, and Doug or I usually landed one every other year while surf casting for fluke. Sand tigers are ovoviviparous, generally bearing only two offspring per litter. They are described as having long, narrow, sharp-pointed teeth and a "trunk crowded with fins of equal size" (Bigelow and Schroeder 1953). Most sand tigers caught in our coastal waters are immature 4- to 6-footers, but big ones are found toward the south-

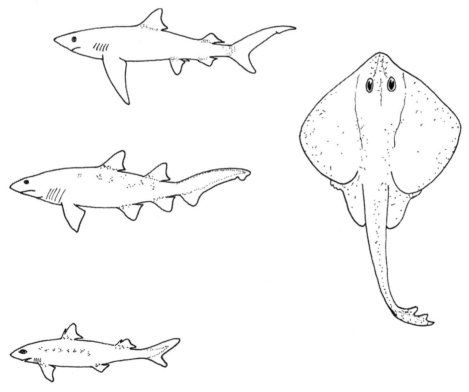

Fig. 4.18. Common cartilaginous fishes of New Jersey: (*top to bottom*) blue shark, sand shark, spiny dogfish, and little skate (*right*).

ern part of the state (the record, an even 246 pounds, was taken in Delaware Bay in 1989).

Doug also caught what we thought then was a "nurse shark" but in retrospect could have been a "sandbar" (a.k.a. "brown") shark of the Carcharhinidae family, to which the smooth dogfish, blue, and tiger sharks belong. The sandbar shark had been the most common and, by the dawn of the nineties, commercially most important shark species in the Mid-Atlantic (Camhi 1998). However, after landings of all coastal species soared from 1991 to 1995, the NMFS concluded that they were being overfished. Thus, given institution of a quota in 1993 and a halving of that quota in 1996, in concert with an odds-on diminution in the sandbar shark population, New Jersey's commercial landings were back to 1989 levels by 1997. Hopefully, the news that brown sharks (and skates and rays) were very abundant in 1998 along the Long Beach Island surf (where they scared a good number of bathers) foretells of better times ahead. On the other hand, Camhi (pers. com. 1998) attributes much of that perceived increase in abundance to a shift in currents that created favorable feeding conditions.

Sandbar sharks are disproportionately more vulnerable to fishing pressure, given that many of their pups are born in Delaware Bay, which they rely upon heavily for summer nursery habitat (Pratt and Merson 1997, cited by Camhi 1998). Great Bay is also an important nursery ground for sandbar sharks. In addition to commercial pressure and/or gill net bycatch mortality, Whitmore (1997, cited by Camhi 1998) estimated that in 1995 nearly 46,000 sandbar sharks were caught by recreational anglers in Delaware Bay, only 69 percent being released alive. The sandbar shark holding the state angling record (160 pounds) was caught in Little Egg Inlet in 1987.

The blue shark (*Prionace glauca*) is a streamlined, intermediate-size pelagic shark. It is one of the more popular of the sharks among sport anglers, and historically thought to be the most abundant of the oceanic sharks, ranging throughout the North Atlantic and into the Mediterranean Sea. Geiser (1969) reported that females about to drop their young dominated the European population, while males were preponderant on the U.S. side of the Atlantic. Liking water temperatures between 67° and 70°F, they are summer visitors, recreational target species, and victims of bycatch by the commercial tuna fleet.

Blue sharks were despised by New England whalers for their habit of gathering to feed on sperm whale carcasses, but they normally feed on fishes smaller than themselves (Bigelow and Schroeder 1953). They also are known for their feeding frenzies and were the subject of Navy tests of repellents (which didn't work) aimed at protecting divers and downed airmen during World War II (Geiser 1969).

Blue sharks are viviparous and the young are 1½ to 2 feet long at birth, 7 to 8 feet at maturity (Bigelow and Schroeder 1953). They have sharp teeth serrated along the margin, and their indigo blue backs and white bellies make them easy to recognize. The biggest one caught in New Jersey was 366 pounds (Mud Hole, 1996). The NMFS includes blue sharks under "other sharks" in its recreational landings database (they only have three categories I could find—dogfish, other sharks, and skates/rays). These statistics indicate a fairly systematic decline in catch from over 100,000 fish a year in 1990 and 1991 to 19,000 in 1996, supporting what Daiber (1995) warned was a worldwide decline in sharks.

Known to just about everybody on the planet now, due to its motion picture infamy, is the great white shark *Carcharodon carcharias*. Well before those flicks, however, New Jerseyans became aware of *Carcharodon* in 1916 when a boy and his would-be rescuer were killed in an unusual invasion of Little Matawan Creek off Raritan Bay in Monmouth County, and two other men were attacked and died just weeks apart in the Spring Lake and Beach Haven surf (Geiser 1969). In 1960 another man lost a leg to a shark of unconfirmed type in the Sea Girt surf, a fact not lost on yours truly as he straddled a long-board or bodysurfed every day from sunrise to sundown at Ocean Beach. Nonetheless, despite logging almost five thousand hours in the surf during my lifetime, no great white, short-fin mako, or tiger shark ever reared its head beside me, and there were only five recorded attacks in New Jersey (none fatal) between 1959 and 1990 versus 54 lightning fatalities over the same time period (Camhi 1998).

Of some 344 known species, these are three of only a handful of sharks interested in or capable of attacking humans (Long 1995), and they do not do that very often. In fact, on a worldwide basis, fewer than one hundred attacks and less than thirty fatalities occur each year (Lane and Comac 1992). The great white is not necessarily the largest shark in our waters either. A 759-pounder holds the state record (off Point Pleasant, 1988), but records for the tiger and mako take top honors at 880 and 856 pounds, respectively (off Cape May in 1988 and in the Wilmington Canyon in 1994).

Other than as bycatch and nuisance potential, sharks have only recently aroused the interest of commercial fisheries in the Northeast, although their livers, which as mentioned earlier are full of oil, were used as a source of vitamin A before that product was successfully synthesized years ago. Now they are being sought as supplies of groundfish (cods, haddock, halibut) dwindle here and demand for cartilaginous tissue soars in Asia, but they are also gaining respect for their reputed medicinal value. That potential was publicized by Lane and Comac (1992) in their book *Sharks Don't Get Cancer*, which promotes a theory of how cartilage may act in preventing or killing tumors. The idea was first hinted at by the fact that cartilage has no blood vessels, and that there must be some good reason for that— i.e., there must be a substance in cartilage that retards vascularization. Since tumors are highly vascularized, the authors theorize that some derivative of cartilage (or a synthetic imitation) might kill tumors by starving them of their blood supply. This theory is very controversial, but the Food and Drug Administration (FDA) has recently bowed to testing it. Should results be encouraging, much more fruitful use could be made of "dead discards" and/or sharks targeted for reprehensible and wasteful "finning" (excising their fins and throwing their bodies away) to satisfy the shark-fin soup market. Based on back-calculations from imports of dried shark fins, Lane and Comac estimate that such finning may kill 5 to 7 million sharks a year (1992).

An NMFS initiative to protect sharks was delivered in an FMP in 1993, but that plan excludes dogfish and skates and applies only to waters under federal jurisdiction (outside 3 miles). Encouragingly, member countries of the United Nations Food and Agricultural Organization agreed in 1998 to prepare (by 2001) an international plan of action to establish sound shark management practices. The NMFS is also in the process of delineating, as part of its mandate under the 1996 amendments to the Magnuson Act, "essential shark habitat" (Camhi 1998). Inshore waters are regulated by the states, however, and New Jersey (with a commercial shark fishery ranking fourth in landings of all species, but principally dogfish) is one of several with no shark management plan at all as of fall 1998 (although as this book went to press the ASMFC and the New Jersey Marine Fisheries Council were working on a plan). Delaware does have a management plan, including a provision that landed sharks must have their fins intact.

The federal plans have been attacked by recreational interests on the basis that they continue to favor commercial interests by allowing a dead discard quota of several hundred tons of blue sharks in the tuna fishery, and by establishing targeted commercial quotas of half a million tons for "pelagics" like mako while proposing to limit the sport quota to one fish per boat. As pointed out in Ristori's 1998/99 columns in both the *Newark Star Ledger* and *The Fisherman*, however, shark quotas continue to cater more to potential socioeconomic impacts on commercial fisheries than to those credited to recreational interests, which in 1984 amounted to an almost $8 million industry, according to Figley (1984, cited by Camhi 1998). As an example of relative impact on shark populations, Ristori highlighted the fact that the coastal harvest of makos by commercial fisheries increased thirteenfold from 1983 through 1994, while the recreational catch fell 22 percent (746,600 to 160,900). That's still a lot of sharks, but the NMFS estimates that only 2 percent are harvested, the rest being released live by New Jersey anglers (who are encouraged to do so, per the "Ristori Amendment" to the FMP, by cutting the leader after the fight is won). I could find no match for New Jersey mako landings in the NMFS landings database, indicating that either their populations have, in fact, been reduced to nil and/or that our own commercial fishermen are not targeting them, and only 924 pounds of tiger shark (two fish?) were landed in 1997 (worth only $121).

The little skate (*Raja erinacea*) gets far less publicity even though it fights well (Bruce disagrees with me on this, saying that catching a skate is like hitching a rock) and is abundant in the New Jersey area. Recreational anglers took more than half a million skates and rays in 1996, up from just 125,000 in 1990, and I would venture to say that most were probably little skates. Little skate are oviparous. They copulate frequently at any time of year and lay those egg cases described earlier. Their young, which are about 4 inches, emerge through a slit at the end of the case and grow to lengths of 8 inches at one to one and a half years, 12 inches at two to three years, and 20 inches by age six to eight, although mortality is high after age five (Bigelow and Schroeder 1953). Little skate are abundant on sandy bottoms during summer and frequently take bait offered for fluke, although they are said to like crabs best.

HIGHLIGHTS OF A TYPICAL "INCIDENTAL CATCH" AT THE SHORE.

In October 1998 the *Newark Star Ledger*'s fishing column announced the taking of what are truly incidental catches for our area, namely cobia (at Normandy), tarpon, barracuda, and wahoo at Barnegat. Normally, however, the incidental (i.e., nontarget) species you are likely to find here most frequently are far less pelagic. I've picked four based on my own experience: the American eel (*Anguilla rostrata*), the northern puffer (*Sphoeroides maculatus*), the oyster toadfish (*Opsanus tau*), and the northern sea robin (*Prionotus carolinus*).

The eel has the most complex life history. It is a catadromous species, doing just the opposite of what striped bass and shad do. It spawns in the sea, but abides in fresh or estuarine waters during its adult life. Specifically, both American and European eels spawn during winter in the Sargasso Sea, which was mentioned earlier relative to bluefin tuna overwintering habitat. This fact was unknown until about seventy years ago when a Danish biologist tracked European eels there and subsequently caught a bunch of unusual-looking fish larvae, which he linked (by a process of elimination, I guess) to the presence of adult eels. Those larvae, subsequently named Leptocephali ("leaf-like"), are transparent, ribbon-shaped forms about an inch long (Lippson and Lippson 1997). Indeed, Izaak Walton talked about the great amount of speculation concerning eel procreation, noting that

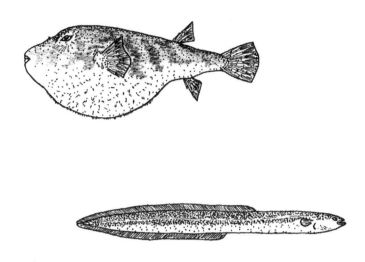

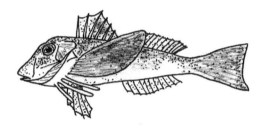

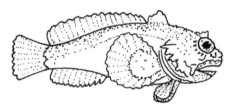

Fig. 4.19. Assortment of fishes you might catch "incidentally": (*top to bottom)*
blowfish (or "puffer"), eel, sea robin, and toadfish ("Sally Growler").

opinions ranged from normal fish reproduction (which no one had witnessed) to the divine (from dew or dust) in his 1653 classic *The Compleat Angler*.

In late February 1969 I collected thousands of eel larvae in midwater trawling exercises that were part of a National Science Foundation–sponsored marine biology training cruise in the Caribbean on the RV *Eastward*, which I was privileged to have been invited aboard. These forms, which drift around at the mercy of currents but somehow are able to know whether they should go to Europe or North America, eventually reach our shores after about one year. When they do arrive in our bays, estuaries, salt marshes, and stream mouths—as early as November but typically not until March—they begin to look more like real eels, with their fin arrangement and pigmentation, and at approximately 3½ inches are called "glass eels" (Lippson and Lippson 1997). Glass eels turn into "elvers," and elvers turn into "whips" at about two years of age, whips then maturing into a "yellow eel" stage (actually, the summer stage is called "green eel," and the fall stage is called "black eel"), which lasts nine to nineteen years for females and four to twelve years for males (Fahay 1995). Males tend to hang out in estuarine waters, but females migrate further upstream into fresh waters, supposedly climbing over dams and other obstructions in order to reach their destination. At maturity, females descend from their freshwater habitat and reacquaint themselves (after perhaps a decade) with their younger male friends en route to their spawning sojourn back in the Sargasso—as "silver eels." At this stage the eels' eyes are bigger, but their internal organs are atrophying (Fahay 1995), probably to conserve energy, for after spawning they will die.

Eels are a common feature of the Jersey shore sport and commercial fishery. Those caught in estuaries like the Manasquan, Shark and Shrewsbury rivers during summer are typically blackish males, and at a size of 2 feet they can be a lot of fun until you get them into the boat. Though eels are more commonly caught in bays and estuaries, Bruce caught a 4-footer in the surf at Monmouth Beach while fishing clams for stripers. Once landed, these slimy, slithering fishes need to be handled with a cloth rag (if you are on a boat) or with hands bap-

Illus. 21. The author hovering over a benthic sampling dredge aboard the research vessel (RV) *Eastward* in February 1969

tized with sand. I am reminded of the day I persuaded Dad to go fluke fishing with me and my friend Mickey in a rowboat we rented out of Manasquan in 1965. I caught a 3-foot eel, and Dad had all he could do to keep from jumping out of the boat as the eel twisted and skipped around the hull. (Dad, for all the opportunities he provided me, was a reluctant fisherman at best.)

Eels support a reasonably successful commercial fishery, given their value as table fare and baitfish for striped bass live-liners (fetching maybe $2 apiece for the latter). Commercial eel landings averaged about 230,000 pounds a year from 1983 through 1997, more than

98 percent of them being taken in "eel pots" in New Jersey, though the percentage being harvested from pound nets increased from zero as recently as 1993 to over 10 percent of the total landings in 1997. The only figure I was able to find for recreational harvests was an estimate of 6,000 pounds in Delaware Bay in 1992 (Fahay 1995). There is a budding controversy over the harvesting of glass eels, which, according to Ristori's December 20, 1998, report in the *Newark Star Ledger*, can draw $350 a pound (maybe 1,000 eels) for export to Japan. Approval was denied by the state senate in 1998, although Fahay's (1995) account of the status of eel populations in the Northeast suggested that stocks had been fairly stable over the previous quarter century.

Eels eat almost anything—dead or alive—and put up a good fight. They are also good to eat, as many Italian Americans know from Christmas dinners. Eel also has its fans among sushi lovers and in Spain, where elvers are prepared and served under their generic name *Anguilles* (Lippson and Lippson 1997). Skinning an eel is accomplished by laying it on its back and making an incision posterior to its head, then turning it inside out (paraphrasing from Geiser 1969). They are usually pickled, smoked, or fried, and the state saltwater record is held by one taken in 1988 at Atlantic City weighing 9 pounds 13 ounces. Interestingly, a 6-pound-2-ounce eel was caught in Round Valley in 1994 (the state freshwater record). The minimum size limit for eel as of 1999 is 6 inches, a size limit that does not factor into the average sport fisherman's catch, but is a hot topic relative to commercial interests and sustainability of the eel population of the Northeast.

The northern puffer, more commonly known as a "blowfish," was once not that "incidental" a part of the bay and ocean catch in the New Jersey area. In fact, it was Mr. Ristori's "first fish" in 1945 (*Newark Star Ledger*, January 3, 1999), predating my first fish by maybe four years. Based on Geiser's (1969) account and my own experience in the late fifties and early sixties, blowfish entered our estuarine waters in such droves that they became the scourge of spring striper and winter flounder fishermen and made the use of bloodworm bait economically prohibitive! Nevertheless, my family loved the taste of these little guys ("sea squab," since they tasted like the chicken-of-the-sea), and Doug and I enjoyed many a day catching them with light tackle and worm bits dangling beneath bobbers in Barnegat Bay off the "telephone pole" landmark south of Toms River aboard Doug's homemade oak and plywood duck boat.

Most puffers are small, the state record holding at 1 pound 14 ounces (again Delaware Bay), but their habit of sucking in water and/or air and inflating their bodies to hard, bloated, little balls makes them a curious study for youngsters and adults alike. As mentioned, they are delicious skinned and fried and, unlike their relatives from the South Atlantic or Pacific Rim, are not toxic unless consumed in amounts of 25 or more per sitting, which might furnish 5 percent of a lethal dose (Geiser 1969). To prepare them, you pretty much follow the same instructions as for eel, except you slit them behind the head without turning them on their backs, and then proceed to turn them inside out, severing their fins as you proceed. Work gloves are a prerequisite for this surgery, given that the blowfish's skin is like heavy-duty sandpaper, and after skinning a bucketful without gloves your hands may resemble bloody sandpaper as well.

The toadfish, a year-round resident of bays and estuaries, has unceremoniously been called a slimy, ragged fish with fleshy flaps and warts (Lippson and Lippson 1997). Equally unflattering is its Navesink River folklore autograph "Sally Growler," a term I heard from a next-door neighbor in Fair Haven during the late seventies, which presumably had something to do with a toadfish attacking (or at least growling at) Sally at some point in history. In fact, toadfish do growl or grunt when pulled out of the water on the end of a hook, as my one-and-only Sally Growler did circa 1979, and during their mating ritual males produce a foghorn sound to entice females into the nest they have readied (Lippson and Lippson 1997). Toadfish have vice-like jaws and stout spines, and the one I caught was essentially brownish black. They have the distinction of producing the biggest eggs relative to their size, up to a quarter inch in diameter. Their eggs, like those of the fathead minnow described in chapter 3, are attached upside down beneath a protective nest of wood, cans, or other debris. They are the only offspring of the saltwater bony fishes included here

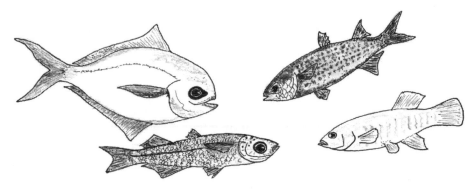

Fig. 4.20. Saltwater baitfishes: (*clockwise from top left*) butterfish, mullet, mummichog ("killie"), and silverside ("spearing").

to be provided parental protection during their formative (three to four weeks from hatching) stages (Lippson and Lippson 1997). Thus their success when loosed into the wider estuarine food web.

Lastly, the poor sea robin. It is, above and beyond any other species you may catch in New Jersey, the most scorned fish of our inshore coastal area. It is abundant, because it is physically too unattractive for anyone to succeed in pioneering a local food fish market (although it was apparently sold in the Hartford area in the 1800s as "wing-fish"). It intercedes in your pursuit of fluke, and it "cusses" back at you (i.e., grunts) when caught (Geiser 1969). The sea robin has an unmistakable appearance, what with its armored head and reddish brown or orangy-colored body. It is common along the surf on sandy bottoms where it may imitate the fluke's behavior of burying itself in the sand with only its eyes and dorsal fin protruding, and it readily hits any kind of molluscan bait you may present. I am telling you about the sea robin only to caution you as to what you may find at the end of your line, even though it and its roe are actually tasty if you can get past the looks!

COMMON BAITFISHES.

Saltwater baitfish common to the New Jersey area include the menhaden (which was discussed in the preceding chapter), the four fishes illustrated here (butterfish, mullet, killifish, and silverside), and two not illustrated since you find them less frequently in bait shops.

The butterfish (*Peprilus triacanthus*), in addition to serving as excellent cut bait for bluefish, has the distinction of not only providing superior table fare but sometimes also offering sport by taking a piece of sandworm bait. Bigelow and Schroeder (1953) describe butterfish as fatty, oily, and having a delicious flavor. They are not commonly found in supermarkets, but I did see some nice ones on ice in a market in Union City in the winter of '98. Butterfish have thin, deep bodies with small and easily loosened scales. They get as big as about a foot in length, though the ones we see in our surf are more commonly two-thirds that size. Butterfish spawn during summer a few miles offshore, and their larvae have the peculiar habit of hiding among the tentacles of jellyfish (Geiser 1969). Due to their protracted spawning season, it is apparently not uncommon to encounter two size groups in the ensuing summer, often wrongly viewed as two distinct year classes.

Butterfish like sandy nearshore areas during summer, and I have seen my share of them flopping around in the surf wash when bluefish were after them. Bait shops typically carry butterfish when they are abundant, and they support a modest commercial fishery in New Jersey. Landings ranged from 268,000 to 1.45 million pounds from 1983 through 1997, peak years (> 1 million pounds) being '85, '86, and '93. Landings for '97 (the last year available to me for this book) were 571,000 pounds.

Though popular in Europe, and found in some of the more cosmopolitan fish markets in New Jersey, mullet is not favored as table fare in our area. The striped mullet (*Mugil cephalus*) is, however, very popular among bluefish and striped bass, and its appearance in the surf in September may be a harbinger of stripers to follow (Geiser 1969). Mullet form large, dense schools that can be recognized as dark pockets of activity beneath the surface. Their backs are dark olive and they are silvery below, and at sizes between 3 and 6 inches they make for a nice whole bait, often hitched to the hook of a "doodlebug," which keeps them suspended over the floor of the sea and in the faces of pelagic species.

Killies take the prize for being the hardiest of our fishes, especially the little "mummichog" (*Fundulus heteroclitus*). Mummichog is an American Indian name meaning "going in crowds" (Lippson and

Lippson 1997), a habit you will find readily observable in any reasonably clear tidal creek or pool. Even if the water is too muddy to
see them, be assured that they are there, for they can withstand some
of the most fouled habitats we have produced over the years. In the
sixties, the mummichog was the only fish species capable of main-
taining permanent spawning populations in the tidal ditches and
creeks of the Hackensack Meadowlands (Quinn 1997), though by
the eighties I also observed carp in a tributary of Berry's Creek behind an office I once staffed in Lyndhurst. Now, although hot spots
of hazardous waste remain in Meadowlands sediments, if you visit the
nature center at DeKorte Park with its backdrop of a landfill in Lyndhurst, you will be in for a pleasant surprise. Two of the reasons why
the mummichog is able to see the turn of the millennium are (1) its
ability to withstand extremely low dissolved oxygen levels in the water, swallowing air at the surface if necessary and (2) the fact that we
put some brakes on marsh "reclamation." Mummichogs are intimately associated with salt marshes, feeding in shallows at the edge
of the marsh, where (among other things) they help control the mosquito population by consuming larvae, and depositing their eggs in
empty mussel shells in the intertidal zone (Smith 1995). "Mummies"
make great fluke bait and are easy to catch with small beach seines
or wire mesh "killie traps." Nonetheless, in spite of their tolerance for
suboptimal environmental conditions, mummichogs may still be susceptible to overharvesting for bait, something I was surprised to learn
when Bruce informed me that bait dealers were having a hard time
keeping up with killie supply and demand in 1998.

Silverside (*Menidia menidia*) are commonly known to the angler as
"spearing," a silvery fish that can be seined during high tide or pur-
chased fresh, frozen, or pickled (depending on the season) at bait-
and-tackle stores. On average, they grow to about 4 inches in length
while living only a year or two, and are common in tidal creeks and
along sandy beaches throughout our bays and inlets (McBride 1995).
Spearing are used primarily as bait, but Geiser (1969) notes that they
can be eaten when prepared like smelt.

The other two baitfish I alluded to earlier are the bay anchovy
(*Anchoa mitchilli*) and the sand eel, or sand lance (*Ammodytes ameri-*

canus). Bay anchovy have wider mouth gapes than spearing and only one dorsal fin, and they are found both in tidal and in open waters (Lippson and Lippson 1997). Their appearance when jumping about to avoid predators so as to dimple the water's surface has led local anglers to refer to them as "rainfish." Sand eels are common to sandy habitats. They are frequently caught with spearing in the same seine, but look quite distinct even though they are silver. They are long, slender, and cylindrical, and unless you examine them closely, their soft skin appears to be scaleless.

The Waterbodies

When the salinity of a waterbody reaches about one-half part of salt per thousand parts water (0.5‰), it is no longer considered to be fresh water. Full-strength seawater is 35‰, and brackish water is within the range between the two. Armantrout (1998) defines brackish water as "water with a salt content greater than freshwater but less than seawater," whereas Reschke (1990) defines water with a salinity ranging from 0.5‰ to 18‰ as brackish. She classifies waters with a salinity higher than 18‰ as being marine.

In any case, all of these waters have two things in common. They have different degrees of saltiness, and they experience tidal changes as a result of the combination of gravitational forces attributable to the earth, moon, and sun. This results in the creation of inter- and subtidal zones, the latter never left high and dry, the former exposed to air between high and low tides, ebbing and flowing in six-hour cycles. In our area, the amplitude of the tidal range is just over four feet on average. When the moon is in either its full (◻) or new (◙) stage, the intertidal zone extends below the "mean low-water" mark and above the "mean high-water" mark. Those are called spring tides. Neap tides, which have the lowest amplitude, occur halfway between those extremes (the moon is half full). The salinity at any given point within tidally affected areas varies as the tide ebbs and floods, as mentioned earlier with respect to striped bass spawning, and organisms must either adjust to this if they can't move (e.g., barnacles), or go with the flow (e.g., fish). Naturally, you cannot expect good fishing off a short pier at low tide, and you would be well advised to get hold of a boat.

Luftglass and Bern (1998) include a variety of saltwater fishing places in their list of 100, and the NJDEP Division of Fish, Game and Wildlife (n.d., but probably ca. 1994) lists the types of facilities that may be found in sixty-eight towns from Sewaren to Salem in its pamphlet entitled *Salt Water Fishing in New Jersey*. The pamphlet provides information, by town, on surf casting, jetty, pier, inlet, boat ramp, and rental facilities, plus general facts about how and when to catch the most popular species. I have extracted those listed for boater access to brackish and/or marine waterbodies:

- Public Ramps: Absecon, Atlantic, Avalon, Barnegat, Barnegat Light, Bass River, Beach Haven, Belmar, Island Heights, Liberty State Park, Keyport, Linwood, Millville, Ocean City, Parkertown, Port Monmouth, Red Bank, Sea Isle City, Seaside Heights, Ship Bottom, Stone Harbor, Toms River, Ventnor, West Creek, and Wildwood.
- Rentals: Atlantic Highlands, Barnegat Light, Bayville, Beach Haven, Belmar, Brigantine, Forked River, Fortesque, Highlands, Keyport, Lavalette, Manasquan, Mantaloking, Matts Landing, Normandy, Oceangate, Ortley, Port Republic, Red Bank, Salem, Seaside Park and Heights, Somers Point, Toms River, Tuckerton, Waretown, and West Creek.

In other words, even if some are too crowded or have come and gone since the mid-nineties, you're never too far from an alternative.

In general, our saltwater resources can be classified as estuarine systems or marine systems, the former varying from brackish to saline, the latter remaining a relatively constant 35‰ Marine habitats can be subdivided into coastal waters over the continental shelf and oceanic waters beyond the shelf. Estuarine systems can be further broken down into rivers, tidal creeks, and bay systems of various types. Able and Kaiser (1994) listed sixteen systems for purposes of compiling their literature search for New Jersey estuaries. They are the Hudson River, Newark Bay (including Hackensack and Passaic rivers), Raritan River and Bay, Sandy Hook Bay (including Navesink and Shrewsbury rivers), Shark River, Manasquan River, Barnegat Bay (including all of the rivers and creeks flowing into it), Mullica River and Great Bay, Brigantine (including Little, Reeds, and Absecon bays), Great Egg Harbor River and Bay (including the Great Egg Harbor and Tuckahoe rivers, plus Lake's and Skull bays), Corson's Inlet (including Ludlum Bay), Townsends Inlet (and Great Sound), Hereford Inlet (including Grassy, Jenkins, and Taylor's sounds), Cape May Inlet (and Jarvis Sound), and Delaware River and Bay and all of their brackish-water tidal tributaries. As illustrated in previous sections, estuaries are very important in the early life of many fish species, which is the subject of a recent book by Able and Fahay (1998).

All of our estuarine systems except the Hudson, Newark Bay, and parts of the Raritan lie in the coastal plain. Most had conspicuous marshes bordering them, due to the submergence of ancient valleys when the ice melted and sea level rose some 250 to 300 feet (NJDRC 1940). These estuaries are therefore said to be of "drowned stream valley origin." Two of the largest of that type are the Hudson and Delaware River estuaries (Reid 1961). The Hudson River is estuarine from Kingston, New York, to its mouth at Bayonne. It supports about 140 fish species and is a major spawning and nursery ground for striped bass, shad, sturgeon, and tomcod. Striped bass and tomcod, plus bluefish, are sought by many fishermen along the Bergen and Hudson County shoreline, whereon numerous bulkheads from Liberty State Park through Palisades Interstate Park provide excellent bank fishing access. Stanne, Panetta, and Forist (1996) provide a complete account of Hudson River ecosystems and fishing in their book *The Hudson: An Illustrated Guide to the Living River.*

Around the other side of Bayonne and up past Jersey City, Union City, and East Rutherford, the Hackensack River also provides good fishing once again. Saltwater species including stripers, white perch, weakfish, snappers, and tomcod invade the Hackensack now that it is cleaner than it was in the mid-1900s, and saltier than it was before the Oradell Dam was completed in 1921. That action effectively reduced freshwater outflow by up to 80 percent, changing the whole ecosystem to a brackish one (Quinn 1997).

Raritan, Lower New York, and Sandy Hook bays all converge in

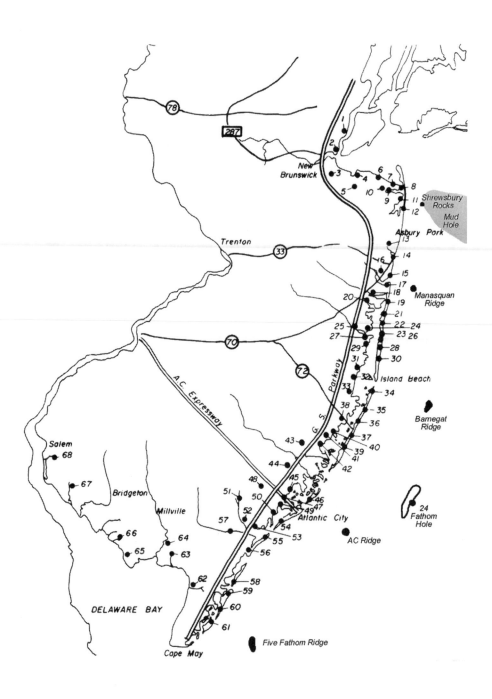

78
287
New
Brunswick
1
2
6 7
4
3 8
5 10 9 11 Shrewsbury
Rocks
12 Mud
Hole
Asbury Park
13
Trenton 33 14
16
15
17
18
Manasquan
20 19 Ridge
21
25 22 24
27 23 26
70 29 28
31 30
72 32 Island Beach
33 34
38
35 Barnegat
36 Ridge
Parkway 37
43 40
39
44 41
42
48
45
51 50
52 46 24
57 49 47 Fathom
54 Atlantic City Hole
Salem 55 53
68 56 AC Ridge
67
Bridgeton
Millville
66 64
65 63
62 58
59
DELAWARE BAY 60
61

Five Fathom Ridge

A.C. Expressway
G.S.
Cape May

the area bounded by Perth Amboy, Sandy Hook, and Coney Island, New York. Raritan Bay has seen its share of insults since hosting as a summer playground for North Jerseyans and Manhattanites in the first half of the 1900s, but it still offers action fishing for fluke, stripers, and other sport species. Sandy Hook Bay provides top-notch fishing. It is fed by the Shrewsbury and Navesink River estuaries, which support almost the entire soft-clam fishery of New Jersey (NJDEP 1998a). That fishery is so important that since 1981 the state has made the watershed a target of some major efforts to reduce non–point source bacterial pollution from, among other things, horse farms. The pier in Fair Haven, created with "Green Acres" funding,

---

Fig. 4.21. *(opposite)* New Jersey's Saltwater Fishing Facilities. Adapted from NJDEP (n.d., ca.1994).

1. Sewaren	24. Island Heights	47. Pleasantville
2. Perth Amboy	25. Toms River	48. Galloway
3. Cliffwood	26. Seaside Heights	49. Northfield
4. Keyport	27. Oceangate	50. Linwood
5. Middletown	28. Seaside Park	51. Mays Landing
6. Port Monmouth	29. Bayville	52. Sculville
7. Atlantic Highlands	30. Island Beach	53. Somers Point
8. Highlands	31. Forked River	54. Ventnor
9. Rumson	32. Waretown	55. Ocean City
10. Red Bank	33. Barnegat	56. Marmora
11. Sea Bright	34. Barnegat Light	57. Tuckahoe
12. Long Branch	35. Harvey Cedars	58. Sea Isle City
13. Neptune	36. Surf City	59. Avalon
14. Belmar	37. Ship Bottom	60. Stone Harbor
15. Manasquan	38. Manahawkin	61. Widwood
16. Brielle	39. Brant Beach	62. Bidwell's Ditch
17. Point Pleasant	40. West Creek	63. Heislerville
18. Metedeconk	41. Parkertown	64. Matts Landing
19. Mantoloking	42. Tuckerton	65. Fortesque
20. Brick Township	43. Bass River	66. Newport
21. Normandy Beach	44. Port Republic	67. Canton
22. Lavellette	45. Absecon	68. Hancock's Bridge
23. Ortley Beach	46. Brigantine	

Illus. 22. A perfect example of the state's efforts to provide boating access to New Jersey's saltwater fisheries and wildlife

also provides a great opportunity for sport fishing. Snapper blues, spot, and winter flounder provide action, and you might even take a Sally Growler, like the one I landed from a canoe in the middle of the Navesink.

The Delaware River estuary is in another classic drowned stream valley. Starting in New York state, the river officially becomes the bay at a point 48 miles upriver from Cape May, just south of Salem. In 1905 that point was marked with a six-foot monument, thus ending a thirty-year dispute between New Jersey and Delaware over fishery and other compacts (*New York Times*, August 9, 1998). Delaware Bay supports a wide variety of fish species, representing almost all of the families described in this chapter. The estuary also hosts spawning runs of striped bass, American shad, and river herrings, the first two faring much better now that water quality has improved, the latter beginning to see more fish ladders around dams on their tidal spawning creeks. From south to north, these include Dennis Creek, West Creek, the Maurice River estuary, Dividing Creek, the Cohansey

River, Stow Creek, Alloway Creek, Salem Creek, Oldmans Creek, Rancocas Creek, and Crosswicks Creek (near Trenton). White perch also spawn in many of these tributaries, the Maurice River fishery, along with its stripers, being the most well publicized.

Barnegat Bay, immortalized in the sixties hit "My Eyes Adore You" by New Jersey's own Four Seasons when they sang ♭♭"Walkin' you home one day, ♫ over Barnegat Bridge and Bay,"♪ is the next great estuarine system in New Jersey. The Intercoastal Waterway, the only deep habitat remaining (and it's not that deep) runs through it, and much of the bay's borders were forever altered in the fifties with housing and lagoon development. The bay is actually one big coastal lagoon behind a barrier beach (coastal lagoon being a scientific term describing its form and creation). It has a high ratio of saltwater volume to freshwater inflow and a low flushing rate, which makes it very susceptible to siltation and nutrient buildup and resuspension by pro-

Illus. 23. Shore fishing access at its best along the Shrewsbury River just north of the Highlands bridge at the base of Sandy Hook

peller action across its shallow profile. Siltation and nutrient-induced phytoplankton blooms can choke the eelgrass beds (beds of long, slender, brown rooted fronds, which provide important habitat for shrimp and other marine invertebrates, not to mention small fishes). An eelgrass blight occurred in the mid-1900s, even before such impacts had been fully realized, and since 1995 Barnegat Bay has been included in the EPA's National Estuarine Program, which provides funding and involvement for implementing a 1993 management plan that focuses especially on reducing runoff of pollutants derived from land.

The northern part of the bay behind the Mantaloking to Seaside Park barrier beach does not support the abundance of fishes it did when I was a youth, but fun fishing for snappers and winter flounder can still be had in the maze of lagoons and channels around its margin, and weakfish also enter the lower end of the bay at Barnegat Light in summer. The book *Barnegat Bay's Fisheries* by Richard Henderson Jr. (check www.online96.com/njfishing) provides colorful and more complete information about the bay and its sport fish species. Great Bay, at the mouth of the Mullica River, was largely spared development because it was too far south of New York City for most commuters. If you take Parkway Exit 58, make a right on Route 9 and a quick left onto Great Bay Boulevard, you'll be able to see and fish the last great expanse of bay and tidal channel habitats in the state. It is a wildlife refuge, but shoreline, boat launch, and boat rental facilities are available.

A good thing that happened in the early seventies was the banning of DDT. That pesticide had been sprayed up and down the streets of Ocean Beach and everywhere else along the shore for mosquito control, and many a kid would get high running behind the twice-weekly "fogging" trucks with their sweet-smelling mist. (Dad always knew better, slamming the cottage windows shut and forbidding such recreation.) Now that DDT has been banned, the ospreys that nested atop every other telephone pole in the forties, but whose eggshells wound up being thin and unprotective after years of parental uptake of DDT through the food chain, have made a comeback in some of our less developed coastal areas.

The intertidal zone of our estuaries encompasses habitats ranging

Illus. 24. The statuesque Barnegat Light guarding the entrance to Barnegat Bay from the ocean

from long tidal flats and shallow networks of channels full of worms, snails, crabs, and mummichogs, to more vertical sub- and partially intertidal environments like piers, rocks, and jetties. The piling illustrated reveals its diverse community of barnacles, mussels, wood-borers, and even sea squirts. Lippson and Lippson (1997) referred to sea

Fig. 4.22. Medley of intertidal pier invertebrates, including barnacles, sea squirts, mussels, and worms. From Alice Jane Lippson and Robert L. Lippson, *Life in the Chesapeake Bay,* 2d ed., 85. Copyright © 1997 by Alice Jane Lippson and the Johns Hopkins University Press.

squirts as "yellowish-green grapes." This type of community varies from year to year through a succession of changes in composition and dominance, and there are hundreds of other invertebrate species in the intertidal and subtidal portions of estuaries that are wonderfully illustrated and described by the Lippsons in their book *Life in the Chesapeake Bay*. Many are found in Barnegat Bay, Great Bay, and the rest of the string of estuaries, bays, and sounds extending from Barnegat Light to Cape May, which feed into the ocean through inlets serving as transitional habitat to jetty or surf-casting environments.

Life in the ocean includes many of the same subtidal invertebrates where the requisite habitat exists, but for the most part the intertidal zone is sandy and subject to powerful and scouring wave action. Fluke, sea robins, skates, dogfish, and the occasional drum dominate the benthic community, while blues, stripers, weakfish, and little tunny are common seasonal members of the pelagic community as they all

Illus. 25. Great Bay in Tuckerton, showing the Atlantic City skyline in the background

Illus. 26. Townsends Inlet in Avalon, providing shore fishing or access to the sea

move in and out with the tides in search of clams, crabs, and sand-worms or pursuing mullet, butterfish, sand eels, anchovy, and silver-sides. Surf casting access can be found most anywhere along our coast from Labor Day to Memorial Day or before nine and after five during summer (these are the best times for fishing, anyway), but Sandy Hook, Island Beach, and some other locations have dedicated surf-fishing areas. The zone just back of the surf provides for excellent troll-ing for bluefish and other pelagic species, or bottom fishing for fluke.

Further out in the coastal zone, blackfish, sea bass, porgy, and cods forage among the artificial reefs, and other pelagics, such as mackerel, the bigger sharks, and bluefin, enter the mix. Further out, oceanic marlin, swordfish, and larger tuna replace the inshore predators of the pelagic fish community, while big cods and pollock inhabit bottom waters. Oceanic fishes rely on phytoplankton as the base of their food chain and oxygen production; most of the species that spawn off New Jersey also produce planktonic eggs and larvae. Coker (1954)

Illus. 27. A prime example of the state's initiatives to provide public access to surf fishing in an otherwise private community on the shore

provides great drawings and discussion of life in the sea in his book *This Great and Wide Sea*.

I added some of the area's popular offshore bottom features, including "Shrewsbury Rocks" and the start of the "Mud Hole," to the NJDEP's map of places to fish featured earlier, but many more interesting ones are found farther out. The drowned bed of the Hudson River is one such feature. Extending southeast from Sandy Hook, the old river channel passes through the Mud Hole, along "Monster Ledge" about 180 feet deep and 35 miles east of Manasquan, and toward the edge of the continental shelf. At the 300-foot-depth contour, it drops precipitously to 600 feet through the "Hudson Canyon," an area of legendary big-game fame, whereupon the shelf ends and the bottom drops more than 5,000 feet in just about 20 miles. These and other natural offshore features are shown and described in the NJDEP's (1982) publication *New Jersey's Recreational and Commercial Ocean Fishing Grounds*. Party and charter boats ply inshore coastal and oceanic areas, and the NJDEP (ca. 1994) prepared a *New*

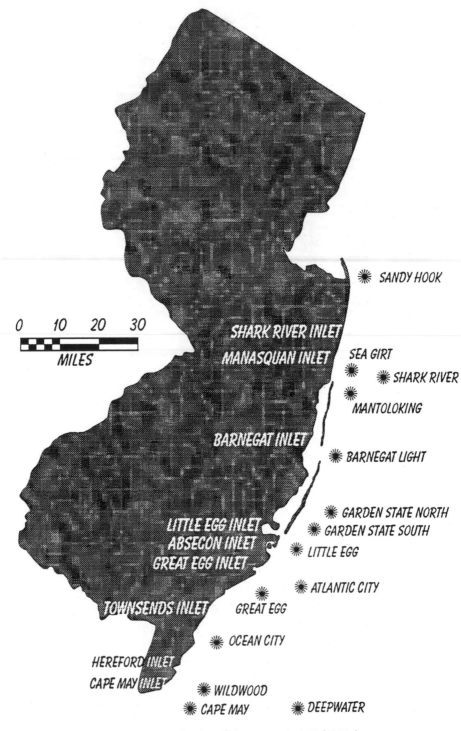

Fig. 4.23. New Jersey's artificial reef sites. From NJDEP (1998e).

Illus. 28. My son Bruce with a beautiful bluefish taken off Island Beach in 1998

Jersey Party and Charter Boat Directory, which lists them by port, boat type, target species/season, and number of passengers that can be accommodated.

Now that you have a rundown of the fishes and their habitats, I'm hoping you will find interest in the remaining chapters on habitat protection and fisheries management approaches.

CHAPTER 5

Factors in Distribution and Abundance of Fishes

O ther than in the Great Lakes, where historical overharvesting by both man and sea lampreys impacted fisheries, inland fishes have been most affected by destruction of physical habitat and fouling of the water. The opposite is true of strictly marine fishes (e.g., the cod), their populations having been hit hardest by overfishing, while water quality remained, until comparatively recently, less of an issue. Anadromous species (shad and striped bass), and species utilizing brackish waters as spawning and/or nursery habitat (e.g., winter flounder and spot) have been subject to all manner of insults and harassment. This chapter sums up the nature of these problems, and how the fisheries scientist examines the extent of them.

The Importance of Habitat

Dr. Seuss's *The Lorax*, a favorite of mine, tells of how things may have been around 1800 when the Union County map shown earlier probably typified New Jersey.

> Way back in the days when the grass was still green
> and the pond was still wet . . .
> From the rippulous pond
> came the comfortable sound
> of the Humming-Fish humming
> while splashing around.
> (from *The Lorax* by Theodore and Audrey Geisel 1971)

174

Unfortunately, as commerce and industrialization burgeoned during the late 1800s, these Humming-Fish found themselves at a serious disadvantage in a rivalry with humankind over rippulous ponds and other aquatic habitat. Proximity to surface waterbodies was of pivotal importance to the import of fuel and raw materials, export of finished products, delivery of a reliable flow of water needed for a variety of purposes, discharge of liquid wastes, and mechanical energy to drive the wheels of mills. As opportunities to build on dry land dwindled, industry turned toward "land reclamation" to satisfy its appetite for waterfront property, and swamps answered the need to remove solid and hazardous waste from our grounds. In the process, tens of thousands of acres of fish and wildlife habitat were lost.

Further fueling this insatiable appetite in New Jersey is the fact that the state enjoys an enviable location for marine transportation of goods to and from anywhere in the world, and is indeed a fulcrum of activity linking New England and New York City with the South and Midwest. New Jersey has some 34,000 miles of roads, which handle 60 billion vehicle-miles a year. The interstate system dates to the fifties, but state-funded highway projects date to 1890, when New Jersey became the first state in the Union to fund such projects. These were done partly in response to lobbying by bicycle riders! In 1909, with automobiles emerging on the scene, we created the first highway commission, and by 1917 had laid out a network of fifteen state highways. Two of these, Old Mine Road in Sussex and Warren counties and River Road in Middlesex County, are in the National Register of Historic Places.

Further expansion of our highway system, a phenomenon everyone expects to be a continuing test of New Jersey's ability to have its cake and eat it too, has also had direct and indirect effects on fish habitat. A two-page spread in the April 26, 1998, edition of the *New York Times* pointed out that engineers and planners are already looking for twenty-first-century, high-tech ways to resolve this dilemma. As the article noted, all a planner had to do in the old days was draw a line on a map and start clearing and grubbing the roadbed — no environmental impact statement, endangered species, wetlands, or any other regulatory hurdles except eminent domain. If a wetland had to be filled, a river channel rerouted, or a creek dammed, so be it.

This next section deals with the environmental aftermath of population sprawl, and some things being done to salvage, enhance, or create new habitat.

ALTERATIONS OF NATURAL WATERSHED FEATURES.

During New Jersey's development, the wetland (marsh, swamp) suffered the greatest damage. Once regarded as wastelands, they are now recognized for their ecological value and their importance in flood control, pollutant filtration, soil erosion and sediment control, groundwater recharge, recreation, research, and for their aesthetic value. But between 1953 and 1973, a peak period of dumping and infrastructure growth up north and housing development down south, we lost almost 62,000 acres of coastal wetlands alone due to filling and diking, about half of that acreage by dredge and fill activity (Zich 1977). Vivid examples of wetland development can be seen in Barnegat Bay, which is surrounded by what started out primarily as summer residences with boat lagoons, and in the Hackensack Meadowlands, within which the New Jersey Turnpike, all railroad beds leading in and out of Hoboken and Manhattan, and Newark Airport's runways sit on top of fill.

In order to stem this tide, New Jersey promulgated its Wetlands Act of 1970 (which deals with tidal wetlands), following this with passage of the Freshwater Wetlands Protection Act of 1988. In 1985 New Jersey still had a total of almost 916,000 acres of wetland distributed over the state's 7,521 square miles, but another 2,551 acres were legally developed from 1985 to 1992 (NJDEP 1998a). Atlantic County has the most acres of wetlands, Cape May the largest percentage, Union the fewest acres, and Hunterdon the lowest percentage within its borders.

Still the largest area of tidal wetlands in northern New Jersey, the Hackensack Meadowlands were created when the last ice plug holding back glacial Lake Hackensack (measuring roughly 50 x 15 miles) melted (Quinn 1997). Despite the loss of two-thirds of its original wetlands acreage, with continued improvement in water quality the NMFS declared the Hackensack River an "Aquatic Resource of Na-

Illus. 29. Wonderful fun for all of us Jersey Coasters in the fifties, but emblematic of a change from wetlands to lagoon habitats prior to enactment of the Wetlands Act. From Cunningham, John T., *The New Jersey Shore*. Copyright © 1958 by Rutgers, The State University. Reprinted by permission of Rutgers University Press

tional Importance," naming a local "Riverkeeper" in the process (the Hudson also has a Riverkeeper, and Delaware Bay a "Baykeeper"). Anyone interested in a thorough, entertaining, and beautifully illustrated account of the Meadowlands should read John Quinn's book *Fields of Sun and Grass: An Artist's Journal of the New Jersey Meadowlands*.

Another of New Jersey's largest glacial lakes was Lake Passaic, through the former bed of which the Passaic River now courses. Lake Passaic was 8 to 10 miles wide by 30 miles long and up to 240 feet deep (NJDCD 1940), but all that remains after the ice melted and deposited its load of solids is a wonderful freshwater wetland, fortunately

Fig. 5.1. A lovely rendering of the Hackensack Meadowlands. From Quinn, John R., *Fields of Sun and Grass: An Artist's Journal of the New Jersey Meadowlands*. Copyright © 1997 by John R. Quinn. Reprinted by permission of Rutgers University Press.

preserved in the Great Swamp National Wildlife Refuge in Morris County.

Although impacts on wetlands receive the press coverage, other forms of habitat alteration affect a great many freshwater fishes. These include destruction of riparian vegetation and stream channelization. Riparian vegetation is defined by Armantrout (1998) as "vegetation growing on or near the banks of a stream or other waterbody that is more dependent on water than is vegetation that is found further upslope." Once removed, soil erosion and overland runoff of rain is exacerbated, leading to siltation in the stream bed, and shade is reduced, allowing temperatures to increase and dissolved-oxygen saturation to decrease. In fact, a totally shaded stream may be 10° or 12°F cooler on a hot summer day than an unshaded stream reach (quoting

Pat Hamilton of the NJDEP in a 1998 internet bulletin on habitat enhancement along the Musconetcong River in Hackettstown).

The *Star Ledger* ran an article on Valentine's Day 1999 about flooding of the lower reaches of the Rahway River. Reporting the effects of that flooding in Linden, Rahway, Carteret, and the Colonia section of Woodbridge, it reads, "A river free of sediment is a river that runs faster. A river that runs faster brings a greater volume of water to the river's end." (The residents were lobbying for dredging to get the water out quicker.) While that may be true now in the downstream reaches, the sole reason that they experience chronic and sometimes devastating flooding is that upstream straightening out of channels (channelization) and destruction of riparian vegetation and floodplains have taken place. The accompanying photo, taken at the skeet shooting range in Kenilworth in 1998, depicts the destruction of the old "horseshoe" where I spent many a day fishing in the fifties. Hopefully, the NJDEP's contemporary emphasis on management of entire watersheds will result in less traumatic experiences for fish and downstream homeowners alike. Evidence of planning ahead for a channel relocation with immediate efforts to restore habitat can be seen in a report by Pat Hamilton concerning a section of Flanders Brook in Morris County (www.state.nj.us/dep/fgw/flanders.htm).

From the fishes' standpoint, soil erosion and resulting waterbody siltation alter the bottom of the waterbody, which obliterates spawning habitat for a variety of fishes needing gravel or pebbles and also has a marked effect on what kind of benthic or epibenthic invertebrates can live there. Channelization, the process of dredging and straightening rivers and streams so as to remove meanders and other forms of natural habitat variety, making them more conduitlike, also reduces biological productivity and diversity (AFS 1971). Fish are especially affected, the larger species moving out due to an absence of pools, shelter, or breeding sites (Hynes 1970). Dams turn tidal marshes into drier types of communities, flowing waters into ponded habitat, and (where no fish ladders or other bypasses are provided) leading to extinction of some runs of anadromous species, as pointed out by Zich (1977).

Illus. 30. Channelization and destruction of riparian habitat at the old "horse-shoe" on the Rahway River on the border of Kenilworth and Cranford in spring 1998

CRITICAL AND ESSENTIAL HABITAT.

The Endangered Species Act of 1973 prohibits the taking of any "listed" species or doing anything to harm their chances of survival, including but not limited to damaging what the act defines as "Critical Habitat." In New Jersey, critical habitat includes those wetlands that are classified as being of "Exceptional Resource Value" (NJDEP 1998a).

Critical habitat should not be confused with what the Magnuson-Stevens Act, the current version of the original Magnuson Act, calls "essential fish habitat," which is defined as "those waters and substrate necessary to fish for spawning, breeding, feeding or growth to maturity," applying to those (saltwater) species the NMFS has responsibility for protecting. This act states that "one of the greatest long-term threats to the viability of commercial and recreational fisheries is the continuing loss of marine, estuarine and other aquatic habitats. . . . Habitat considerations should receive increased attention for the conservation and management of fishery resources of the United States" (AFS, American Oceans Campaign, and NOAA, National Oceanographic and Atmospheric Adminstration, 1998). To that end, FMPs must be amended to define and identify essential habitat; identify adverse impacts, including but not limited to those related to the fishing process itself (e.g., dragging); and develop conservation and enhancement measures.

As described in chapter 4, the New Jersey Division of Fish, Game and Wildlife has been implementing one measure of habitat enhancement for some time now, that being their artificial reef program. The objective of that program is "to construct hard substrate reef habitat in the ocean for certain species of fish and shellfish, new fishing grounds for anglers, and underwater structures for SCUBA divers" (NJDEP 1998e). To determine how well the reefs may be expected to attain that objective, New Jersey biologists employ experimental habitat units, thirty of which were deployed in October 1996. They are just one form of the artificial substrate array that has been used for decades to study everything from effects of water quality on aquatic communities to quantifying the extent of barnacle and mussel encrustation on cooling-water intake pipes or oil rig underpinnings ("biofouling"). The ones used for reef study employ a wire mesh cage set into a concrete-filled tire, packed with layers of corrugated fiberglass and conch shells and topped with panels of varied products. The aim is to find out which organisms settle on what substrates, how communities vary over time, and how then to optimize design and location of artificial reefs.

More information on essential fish habitat may be obtained by checking the web sites www.nmfs.gov and www.americanoceans.org.

OTHER MAN-MADE HABITATS.

There are a fair number of things that can be done to create, re-create, or enhance habitats. A whole field is now devoted to this subject, called "ecological engineering." By any other name, ecological engineering has been around for ages, starting with pond culture, then design of fish ladders anadromous species can scale to get around dams, and a variety of contraptions to enhance stream habitat. Restoration of wetlands is only the latest application of ecological engineering.

REEF COLONIZATION STUDY UNIT

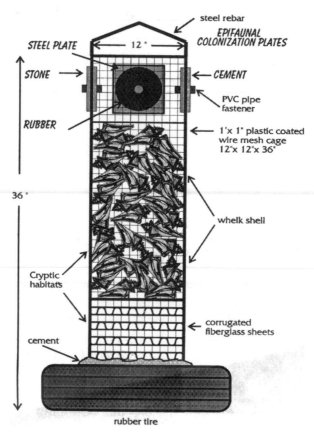

Fig. 5.2. Reef colonization study unit. From NJDEP (1998e).

In 1976 Woodhouse, Seneca, and Broome wrote a report about propagation and use of *Spartina alterniflora* (saltmarsh cordgrass), one of the most important species of low tidal marsh plants, as a means of abating shoreline erosion in estuaries and bays. In 1998 the Army Corps of Engineers embarked on a massive undertaking to reestablish New Jersey's ocean beaches with something other than jetties and annual sand replenishment. The technique they employed, up and down the coast, was one involving construction of a fortified berm parallel to shore, and re-creation of artificial dunes with a representative assemblage of stabilizing native grasses. Creation or restoration of wetlands is also now being done to "offset" (trade off) unavoidable impacts in one place or of one kind with improvements of an equal or greater extent in another. Tidal wetland restoration as a condition of operating the power plant in Salem is one example, and creation of freshwater wetlands—with their assortment of cattails, arrowhead, pickerelweed, and other pondweeds—has been made part of "remedial action plans" for restoration of former hazardous waste lagoons.

Construction of ponds for ornamental, irrigation, or fishing purposes is an old science—and art. Farm ponds, the likes of which New Jersey's two record-size sunfish were taken from, are often not just accidental features of the landscape. Location, design, construction, and maintenance relative to the species you hope to propagate are all critical to the success of the pond. Some people want trout, but whether or not that goal is realistic depends on whether the water is deep enough to remain below 74°F for most of the summer (Eipper and Regier n.d.). Likewise, the success of largemouth bass (the usual recommended warm-water species) depends on whether the surface water remains *above* 72°F in the summer, and the water is deep enough not to freeze completely or become deoxygenated under snow-covered ice, which limits light penetration and plant photosynthesis. I have not made a survey of farm ponds in New Jersey, but I would expect that largemouth bass–bluegill and bass–golden shiner combinations are most common. Bass-bluegill communities like the one described in chapter 3 are initially stocked with ratios of 10:1, bluegill:bass fingerlings (each about 1 to 2 inches long). Beyond fish types and numbers, Eipper and Regier (n.d.) highlight the more subtle, but certainly not insignificant, elements of farm pond maintenance and design, like muskrat (burrow) control, weed control, not locating your pond at the lowest part of a watershed whose area drains more water than the dike can retain, and not setting the pond below a barnyard drainage area!

Construction of fish ladders may be considered a form of stream improvement, at least as far as the anadromous fish is concerned. Like pond design, fishway design requires substantial knowledge of fish behavior, hydrology, and engineering. A fishway consists of a water-filled lock, channel, or series of connected pools enabling the fish to swim past the obstruction. The flume into Deal Lake is one example. Others include the Elizabethtown Water Company's fishway on the Millstone River in Zarephath (Somerset County) and one funded by PSE&G into Sunset Lake on the Cooper River in Camden County.

Other forms of stream habitat enhancement produce better fishing by providing more shelter or diversity, examples being current de-

Illus. 31. Habitat and shoreline protection enhancement at Sea Bright in 1998

Illus. 32. The Deal Flume, a fishway at the shore providing access for river herrings to spawn in Deal Lake

flectors, anchored trees or brush, or cross-stream "digger logs" that create pools below them. Lagler (1956), Rounsefell and Everhart (1953), and Armantrout (1998) provide illustrations of the variety of stream enhancement devices available to the fishery biologist, and one form (the "wing dam") can be viewed along with riparian habitat improvements on the Musconetcong River where it runs past the cemetery in Hackettstown.

How Clean Is Clean?

As time passed, the Humming-Fish also had to contend with water quality changes.

> You're glumping the pond where the Humming-Fish hummed!
> No more can they hum, for their gills are all gummed.

So I'm sending them off. Oh, their future is dreary.
They'll walk on their fins and get woefully weary
in search of some water that isn't so smeary.
    (from *The Lorax*, by Theodore and Audrey Geisel 1971)

The Geisels must have been prophets, for their book was written before tales of "walking catfish" began to surface in Florida. Walking catfish, another exotic import, were said to have become such a menace that people there were reporting incidents of the fish attacking kids and dogs. Walking catfish apparently take to land for reasons other than pollution, even though that is a useful survival tactic, but nonetheless, Dr. Seuss's point is well taken.

The subject of how clean clean is came to prominence in the eighties with regard to cleanup of hazardous waste sites, but the question has been argued for years relative to site-specific limits on concentrations of pollutants released from effluent discharge pipes. At some level the discharge is clearly too dirty (smells, slimes, fish kills, no fish left), but is the cost of equipment, operation and maintenance, and the energy needed to reduce contaminant levels to zero worth the price? That's often the question when one starts looking, on a waterbody-specific basis, at the benefits (B) versus the costs (C) of incremental reductions in concentrations toward the cleaner end of the scale. The big leaps from filthy to something that seems to look clear and doesn't smell or result in fish kills are seen to result in obviously positive B/C ratios. Those dramatic reductions have also typically been the easiest and most efficient to accomplish with a limited number of processes (a short "process-treatment train"). Further cleanup then requires more sophisticated, generally less efficient measures (from an energy to percent reduction standpoint). Recently, emphasis on treatment has been traded for waste minimization or "pollution prevention" efforts, which may include substitution of chemicals that accomplish the same purpose but are less toxic, or recycling of forms that may have intrinsic value to another kind of user (for sale to them) or can simply be used again and again (e.g., solvents).

Determining what concentrations of contaminants are detrimental to individual species and/or may be appropriate to each point source and waterbody combination is the job of biologists, ecologists, toxicologists, statisticians, hydrologists, and (not infrequently) lawyers. Engineers and cost estimators then have to determine how to achieve the desired results in the most cost-effective manner. Pollution has long been detected and measured by chemical tests of the water, presence of "indicator organisms," or effects on the community as a whole (Odum 1959). The following section provides a sense of the underlying principles of those methods, typical tests on live (at the start) organisms exposed to a range of parameters, and how the information is being used to answer the question "How clean is clean?" before the Humming-Fish exit stage left.

BASIC BIOLOGICAL PRINCIPLES.

Surface waters contain a potpourri of gases (oxygen, carbon dioxide, nitrogen, methane), minerals (calcium, sodium, potassium, iron, etc.), soluble organics (tannin, humic acids, proteins), and suspended inorganics (clay and soil particles) derived from geologic formations; riparian vegetation; the plants, animals, and microorganisms living in the water; and the atmosphere. The balance and interactions among these constituents, which change with hydrology and temperature, create an environment that a fish species either likes, tolerates, or can't stand. Dissolved-oxygen (DO) concentrations and pH are two examples, both being affected by temperature, the balance among photosynthetic (oxygen-producing) and respiratory (oxygen-depleting) processes, and salinity. The low end of the DO scale and both ends of the pH scale (too acidic or too alkaline) affect fish distribution and abundance. Add to that mix compounds of human, or "anthropogenic," origin (oil, pesticides, herbicides, toxic wastes), and the suitability of the fish's habitat relative to its physiological needs is further modified.

One of the most important principles of ecology in relation to water pollution is Shelford's "law of tolerance," which he put forth in 1913. As described by Odum (1959), it includes the following tenets:

1. All organisms have an ecological minimum and maximum; too little or too much of something can be limiting.
2. Organisms may have a wide range of tolerance for one thing and a narrow range of tolerance for another.

3. Those organisms with wide ranges of tolerance for everything are likely to be most widely distributed.

4. When conditions are suboptimal for one factor, their tolerance for another may be reduced.

5. These limits vary geographically and seasonally.

6. Sometimes an organism is found not to be living in its optimal range for a physical factor, in which case some other factor is taking precedence.

7. The period of reproduction is usually a critical period relative to an organism's sensitivity to environmental factors.

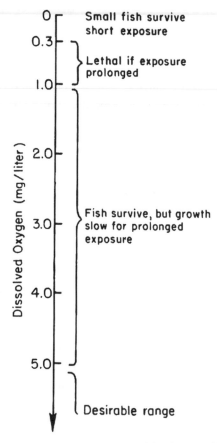

Fig. 5.3. Chart showing oxygen tolerances of freshwater fishes. From Claude Boyd, *Water Quality in Warmwater Fish Ponds* (1979), 71. Reprinted courtesy of Claude Boyd.

Rule 2 is most important in determining what species are to be found in a particular waterbody, or which species may be the best ones to select for toxicity tests. For example, an organism with a narrow range of tolerance for temperature would not be expected to be found in a small mid-Jersey pond, nor would one with a wide range of tolerance for pH be a good choice on which to base effluent limitations for that parameter. Rule 4 is important to remember because tolerance for toxics may be reduced if an organism is forced to live outside its optimal temperature and pH ranges simultaneously. A fish's tolerance for some variables, such as pH, DO, temperature, or whole effluent, can be deduced from field measurements and observations, while lab toxicity tests are essential to developing water quality criteria and sometimes monitoring discharge compliance.

PROTECTING FISH FROM WATER POLLUTION.

New Jersey's water quality standards are the foundation of our water resources management (NJDEP 1998a). The federal Water Quality Act of 1965, followed by the Water Pollution Control Act of 1972 as amended in 1977 to become known as the Clean Water Act (CWA), required the states to adopt standards to protect public health and welfare while considering different uses of each waterbody (including fishing). Water quality standards constitute a package of regulations that include policies, waterbody use designations (e.g., drinking, irrigation, recreation), and criteria. Water quality criteria are comprised of narrative and numerical recommendations for ambient water quality, the former stating goals, the latter being scientifically derived concentrations of pollutants and the limits required for protection of human health and aquatic life (USEPA 1991). The CWA was amended again by the Water Quality Act of 1987, which demanded even greater emphasis on toxics control.

Water quality criteria attempt to consider a wide range of toxic "endpoints" of a test procedure, including death (acute toxicity), reduced growth, or lessened reproduction (subacute or chronic toxicity). The test procedure is called a *bioassay*, which is "a procedure in which the responses of aquatic organisms are used to detect or measure the presence or effect of one or more substances, wastes, or environmental factors, alone or in combination" (APHA, AWWA, and

WPCF1980). Bioassay is a complicated science, if not an art, requiring rigorous quality control from beginning to end. Strict standards for water supply; test fish; experimental conditions; and measurement frequency, accuracy, and precision must be followed. Boiling it down to its bare essentials, however, a representative acute bioassay involves filling a number of containers with a series of concentrations of the solution being tested ranging from 100 percent test solution to 100 percent clean water (the "control"); placing an equal number of fish into each container; checking them at prescribed intervals to count and remove dead animals; and using the mortality and analytical chemistry data to calculate what concentration would be estimated to kill, say, 50 percent of a group of test fish, that being one form of toxic endpoint, i.e., the LC (lethal concentration)$_{50}$. There are a variety of slants on these lab tests, some involving relatively short-term experiments without refreshing the test waters ("static bioassays") and some involving four-day exposures with constantly refreshed baths ("flow-through bioassays"). Further, although my focus has been on fish, similar tests are also performed using algae and invertebrates (water fleas, small shrimp, insect larvae) in order to get a complete picture of potential ecosystem impacts.

The LC_{50} has frequently been used as the first step in figuring water quality criteria or effluent limitations, subsequent steps for fulfilling either objective involving statistical or mathematical manipulations of the data too elaborate to illustrate here. Confirmation that these objectives are being met or attained by the discharger or in the waterbody is obtained by monitoring either or both using physical-chemical tests, more bioassays, or inspection of the benthic invertebrate community at regular intervals. The United States Geological Survey maintains a number of permanent automated stations that monitor water quality in rivers and lakes around the nation. These provide useful data on watershed quality, not necessarily related only to point sources of contaminants.

Non–point source (NPS) contamination—with its load of solids, pesticides, herbicides, bacteria, lead, zinc, mercury, and oil—is now recognized as being on a par with end-of-the-pipe pollution in terms of impact potential. This was beginning to be recognized

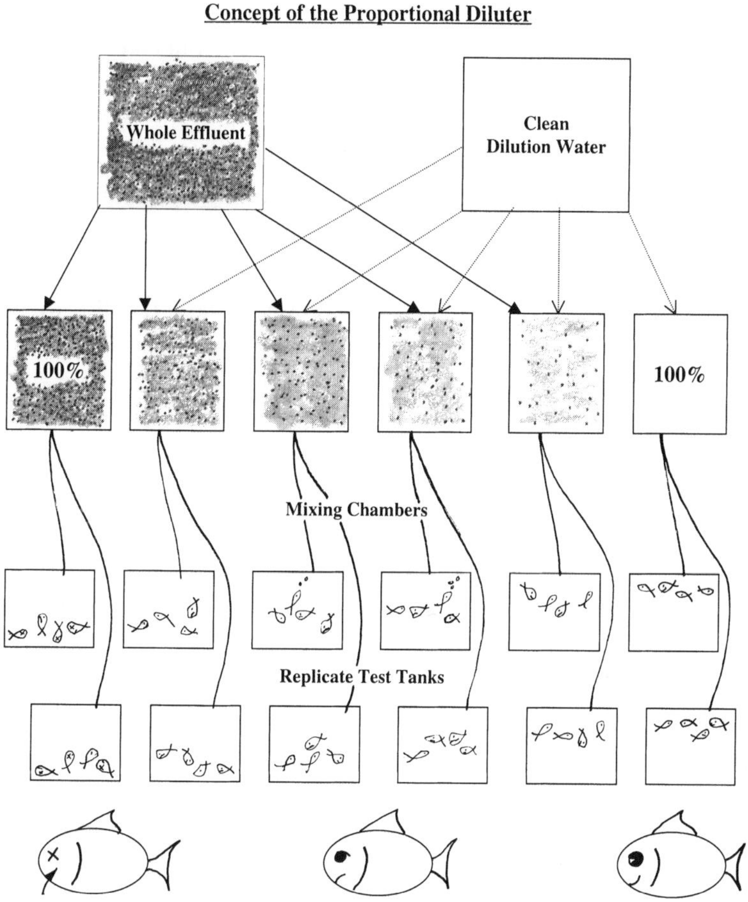

Concept of the Proportional Diluter

Fig. 5.4. Schematic of the proportional diluter device for testing a fish's tolerance for toxics.

twenty-five years ago, with findings that urban storm-water runoff compared in quality with effluent from a municipal wastewater treatment plant (Lager et al. 1977). When NPS or point source pollutants find their way into a body of water, they not only degrade the water but foul the bottom upon settling out. Up for consideration by Congress in the new millennium is a proposed amendment to the CWA called the Fishable Waters Act. This would authorize the states to establish "watershed councils" that address NPS and habitat issues. Another source of contamination in bottom sediments is boats, even private fleets berthed in marinas throughout our lakes and brackish waters.

Sediment quantity and quality have immediate effects on the benthos and are cause for concern down the road when the waterbody has to be dredged. One of the biggest problems facing New Jersey, New York, Pennsylvania, and Delaware is how to continue to operate (let alone expand) the economically important ports on the Delaware and Hudson rivers and the Newark Bay complex. At the "Business of Dredging Conference" held in Somerset in 1996, New Jersey Maritime Resources (NJMR 1996) pointed out that there are about eighty ports and terminals and four hundred marinas scattered throughout our navigable coastal waterways. They are of immense value to commerce, industry, and recreation. In terms of commerce, it has been estimated that one-tenth of the nation's cargo (850 million tons) spends some time on New Jersey's roads, and the U.S. Department of Commerce is seeking to expand New Jersey's share of the port business. This is not possible at present due to requirements for channel berthing depths and increased maintenance and the fact that new dredging must be done over and above the ≈ 62 million cubic yards dredged each year from 1983 through 1993. The problem is what to do with the dredge spoils. They can't continue to be dumped offshore due to contamination by substances listed earlier plus PCBs and in some cases dioxin, but those same contaminants also limit other options.

To solve this dilemma, the Port Authorities on both sides of New Jersey, the EPA, the state environmental agencies, the Army Corps of Engineers, universities, private corporations, and environmental groups have been trying to work toward mutually acceptable alternatives. Options range from creation of containment islands or capped subaqueous pits in the area bounded by Sandy Hook, Staten Island, and Jones Beach to more broadly beneficial solutions. Possibilities being evaluated are redevelopment of former industrial sites termed "brownfields," creation of reefs and wetlands, beach replenishment, landfill cover, and acid mine reclamation—all depending on the quality of the sediments, which varies. As this book went to press the EPA had proposed a set of chemical-specific sediment quality criteria, which had been some time in the making. However, they and the U.S. Army Corps of Engineers, which has responsibility for dredging, were still immersed in a long-running dispute over test procedures and whether the EPA should use the term "guidelines" instead of criteria.

INDICATORS OF CHANGE.

Shelford's law of tolerance has long been the basis for pollution evaluations, particularly those related to domestic wastewater, which, if left untreated as was the case in the first half of the 1900s, eats up oxygen in the water and creates septic conditions. Odum (1959) described the classical succession of organisms above, in, and below a wastewater outfall—all a result of different species' tolerances. Above the outfall (discharge conduit) the water is clear and carries a lot of oxygen, and the habitat supports an assemblage of green algae, diatoms, mayflies, stone flies, caddis flies, and game fishes. At the point of discharge the water is turbid; the oxygen level starts dropping; and fungi, paramecia (forms of tolerant protozoa), midge larvae, carp, and catfish dominate. A short distance below the outfall there is no oxygen; there are no fish; and blue-green algae (which have signature odors), worms, and mosquito larvae are all that may be found. The sequence reverses itself with further distance downstream. Each of the species mentioned, because of their characteristic association with different zones, came to be known as "indicator" species. One must be careful, however, not to confuse lack of physical habitat with pollution, especially in the case of benthic invertebrates.

Other indicators, or indices, of aquatic ecosystem health that have been used over the years are species diversity, redundancy and richness (mathematical constructs using data on numbers of each species present in a sample), and reductions in size-at-age group. In the nineties, biologists started expanding these notions into biocriteria and multimetric bioassessments, introducing acronyms like "RBPs" (rapid bioassessment protocols), "BCIs" (biotic condition indices), "ICIs" (invertebrate community indices), and benthic or fish "IBIs" (index of biotic integrity). The thinking here is that the communities integrate all of the variables in their environment, and thus multimetric indices have a detection capability over a broader range of "stressors" than, say, single-species toxicity tests (Stoughton 1997). The major criticism of these indices revolves around statistical considerations, but nevertheless, in 1989 the EPA mandated an "integrated approach" to water quality–based toxics control using a "triad" of chemical-specific tests, whole effluent bioassays, and bioassessments (USEPA 1991).

Indicators of aquatic quality may also be seen on or in the fish themselves. They include (from Lagler 1956) such afflictions as "furunculosis" (boils and internal hemorrhaging, with superficial ulcers), "water mold" (tufts of gray fungal filaments on the body), and "fin rot" (disintegration of the skin between the fin rays). Fin rot on winter flounder has been observed in, among other places, Sandy Hook Bay and even relatively undefiled Great Bay (Able and Kaiser 1994). To the list of maladies can now be added a new form of disease identified in Chesapeake Bay tributaries, *Pfiesteria*, a "dinoflagellate" officially classified as a type of alga but more of a "missing link" between plant and animal kingdoms. According to a 1998 report from the University of Maryland, which has a web site devoted to fish health (www. mdsq.umd.edu/fish-health), the organism appears to use a toxin to stun fish, then feeds off them while also causing external pink lesions. On September 28, 1997, the *New York Times* reported that the National Institutes of Environmental Health Sciences in Washington, D.C., had documented a few cases of people reporting problems ranging from rashes to memory loss after eating infected fish.

The best advice is to not eat fish from polluted waters, waterbodies that have had reports of fish kills, or fish that just appear or behave as though they are ill, although the state has also issued consumption advisories for fish that have accumulated things that don't make them look or act sick. Those things are PCBs in striped bass, white perch, eel, and white catfish (not channel cats) in many estuarine waters proximal to industrialized areas, and mercury in pickerel and largemouth bass in fresh waters. The source of that mercury is not today's point source effluents, but atmospheric fallout (including concentrations in snowfall) from smokestack emissions, weathering of some geological formations, old dumps (batteries, thermometers, paints — i.e., things most people just throw out in the garbage), and pesticides (NJDEP 1998a). Most warnings are directed toward the "at-risk" category of people (pregnant women or children under the age of five),

194

the warning being to not eat more than one fish per week from some waterbodies, and none from others.

Fishing for Fun or for a Living

According to an account I read in issue no. 5 of the *New Jersey Fish-Net* dated July 9th, 1997 (www.fishingnj.org/njnet5.htm), 16 million people participate in sport fishing in our nation's estuaries and oceans, whereas 240 million do not. The point being that those who *don't* fish, but still enjoy eating fish in restaurants, should not be denied the privilege. That web site is sponsored by the Cape May Seafood Producers Association, the Family and Friends of Commercial Fishermen, the Fisherman's Dock Association, Lund's Fishery, the National Fisheries Institute, and Viking Village Dock. On the opposite side of the coin, Pete Barrett's response to a letter to the editor in the June 13, 1996, edition of *The Fisherman* noted that, from an economic standpoint, saltwater sport fisheries create roughly ten times the number of jobs in New Jersey (9,915 in tackle shops and marinas vs. 1,140 in commercial jobs), and proportionally greater sales revenues ($326 million in tackle sales vs. $22 million in commercial finfish sales).

That's probably in the ballpark relative to jobs and sales directly related to catching and selling finfish. Statistics downloaded from the NMFS web site indicate that New Jersey's commercial fishermen grossed about $92 million a year from 1988 through 1997, most of those revenues deriving from shellfish dredge fisheries. In 1997 surf clams brought in roughly $27 million, quahogs ("cherrystones") $14 million, scallops ($13 million), and oysters $2 million. Pot and trap fisheries took in about $4 million worth of eels and $3 million each of blue crabs and lobsters. Fluke and monkfish (bottom trawls), mackerel and porgies (gill nets), menhaden (purse seines), and tuna/swordfish (baited long-lines), the dominant commercial species in terms of their dockside value, grossed between $2 and $4 million each per year.

The employment figure (1,140 jobs) for commercial fisheries does not factor in the many thousands of part- and full-time jobs created for fast-food clerks, waiters and waitresses, dishwashers, short-order cooks, chefs, and fishmongers, who make their livings serving both anglers and nonanglers in our fresh-fish markets, seafood houses, diners, and fast-food joints that rely on the commercial catch. Similarly, however, figures quoted for jobs and sales in the recreational industry do not include expenditures on dining, lodging, or transportation on-site or en route to the shore (costs associated with "the whole recreational experience") that are incurred by people whose main reason for visiting the coast is to fish. The estimate given in my introduction to chapter 4 ($1.5 billion) does.

The bottom line is that fish are important to most everyone, and in cases where the landings and biological data demonstrate an obligation to restrict harvests of some species, commercial fishermen and sport anglers alike may be required to accept unpopular regulations. Sharks and tunas are only the latest examples of fishes in need of greater protection from commercial overharvesting, but to dismiss the contribution commercial fishermen have made to New Jersey's coastal heritage carte blanche would be unjust, especially considering the irreverently excessive harvests of offshore species by foreign fleets. In his reply to the previously mentioned letter to *The Fisherman*, Mr. Barrett concluded that "only a few wild-eyed radicals want all commercial fishing banned." By the same token, anglers are dealt an injustice if they are banned from keeping their catch, certainly if they intend to keep only that which family and friends can eat. My sons, Bruce and Geoff, practice "CPR" (in this case catch/photo/release), a doctrine they find very rewarding. Here is how the scientist sees the job of fishery management.

WHO OWNS THE FISH?

"A BASIC characteristic of all fisheries is that they are common property natural resources" (Christy and Scott 1965). As of 1994, in America, that translates to all 270 million of us owning the fish (Wallace et al. 1994).

In days of old, when people believed that the sea's resources flowed inexhaustibly from a "cornucopia," entry into a career of fishing was limited only by one's ability to build or buy a vessel and some nets, and hire a crew that would fish as tenaciously as possible to catch and land as many fish as they could before the next guy did. As a supply

Illus. 33. The commercial fishing fleet opposite Ken's Landing alongside the Route 35 bridge in Point Pleasant Beach

dwindled, less well-heeled enterprises would either switch target species or drop out of the business. Wallace et al. (1994) recall the story of the "Tragedy of the Commons," that being coined in England where nobody took responsibility for grassy areas in the public domain, which were known as "commons." As people kept putting more and more sheep on the commons, they were soon overgrazed and, tragically, lost to everyone.

Historically, ownership of the seas extended only so far as a cannonball was supposedly able to fly (3 miles). This was first suggested in 1703 by a Dutch juror; formally recognized by an Italian in 1782; picked up by the United States in 1793; and adopted, in practice or by treaty, by most nations in the nineteenth century (Christy and Scott 1965). With the accession of larger and more sophisticated fishing vessels able to remain at sea longer and locate, hold, and preserve more fish, in tandem with expansion of the nation's interest in oil and greater range of defenses, the limits were pushed. The "Truman Proclamation" of 1945 (the same year the United Nations was formed) extended control, including regulation of fisheries, to the limits of the continental shelf, and in 1964 Congress banned fishing in those waters. Enactment of the Submerged Lands Act in 1953 established the states' control within the 3-mile limit, and the zone extending from 3 to 200 miles offshore, in which foreign fishing is now prohibited, is sometimes referred to as the "Exclusive Economic Zone," or EEZ (Wallace et al. 1994).

A CORNUCOPIA THEY'RE NOT.

Fishermen are just another form of predator, but one that is rarely eaten by anything else and is very potent relative to any other omnivore (something that eats both plants and animals). Other than periodic extremes of climatic conditions, we represent the only agents capable of altering the composition of fish communities and populations, sometimes to the point where individuals cannot replace themselves fast enough to keep pace with population mortality rates. The goal of the fishery biologist is to estimate the number of fishes that can safely be removed by fishermen while still ensuring that the fish population remains healthy (Wallace et al. 1994).

Achievement of this goal entails a three-pronged approach involving biologists, statisticians, and mathematical modelers. The biologist, using samples of fish collected in scientific surveys or taken from commercial catches, examines the age and sex composition of the population; the ages and sizes at which the fish mature; how many eggs the females produce at each age ("fecundity"); and their migratory, spawning, and life stage–specific distributional preferences and feeding habits. The New Jersey Division of Fish, Game and Wildlife conducts trawl surveys five times a year over an 1,800-square-mile area of the ocean from Sandy Hook to Delaware Bay in order to obtain some of these data. The data are then pieced together by the biologist and statistician ("biometrician" in the case where they are one and the same person) to produce growth and survival curves and equations that are essential to understanding the population and evaluating management strategies.

Age data, which are based (as in forestry with tree trunks) on how many growth rings are counted on hard parts such as scales, otoliths (ear bones), or spines, are fundamental to establishing when the fish

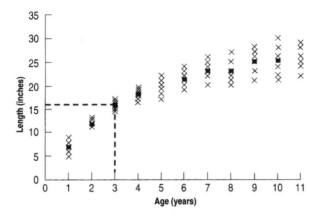

Fig. 5.5. Length-at-Age Key, relating fish size to age. Courtesy of R. K. Wallace, W. Hosking, and S. T. Szedlmayer. 1994. *Fisheries Management for Fishermen: A Manual for Helping Fishermen Understand the Federal Management Process.* Auburn University Marine Extension and Research Center, Mobile, Alabama.

first spawn, how old they may get, and how fast they get there. Size-at-age keys are important tools for determining growth rates as well as mesh-size restrictions for commercial nets (the smaller the mesh size, the smaller and younger the fish has to be in order to escape, or, failing that, be "recruited" into the fishery). If the size and age at recruitment is at or below that which the fish must attain before it can spawn, and too many are taken, the fishery may experience "recruitment overfishing" (Wallace et al. 1994).

Age-composition data, catch-per-unit-of-effort (CPUE, or C/E) data, and mark-recapture study data are also used to estimate mortality rates of fishes and their population size. All rely on ratios of numbers from the start of some benchmark to numbers observed at the end of a specified period. CPUE is an index of abundance and population trends calculated by dividing the numbers or pounds of fish caught with a particular type of gear by the number of units of effort used in making that catch. Weight per one-hour tow of a net, weight per 100 feet of overnight gill net sets, and numbers of fish per angler-hour are examples. A reduction in CPUE, rather than just landings, is a sure sign of trouble (Wallace et al. 1994).

Skipping the elaborate details here, basically a mathematician is pressed into service to roll all of the biological parameters up into

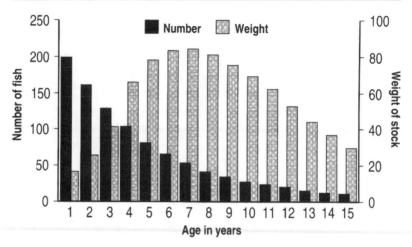

Fig. 5.6. Relationship of weight and number of fish in a stock. Courtesy of R. K. Wallace, W. Hosking, and S. T. Szedlmayer. 1994. *Fisheries Management for Fishermen: A Manual for Helping Fishermen Understand the Federal Management Process.* Auburn University Marine Extension and Research Center, Mobile, Alabama.

some models of numbers and weights of fish in the population, by age group, and model the fishery. If *numbers* of fish per age and their combined *weight* per each age are plotted, one observes that, while numbers decline with each advancing age, the total weight of the population swells through growth of survivors until even the oldest ones start succumbing to fishing or natural mortality. It is clear from this picture that even though a fishery might avoid recruitment overfishing, it may be taking the majority of fish before they can realize their potential in terms of weight for the fishery. This is called "growth overfishing" (Wallace et al. 1994). Fishery regulations typically focus on postponing the age of recruitment as long as is practicable, usually by establishing minimum size limits, and dictating quotas for the numbers (sport fisheries) or weights (commercial fisheries) that may be taken per day, season, or year, sometimes shutting the fishery down completely to protect spawners.

Enter the bycatch dilemma. In the Magnuson-Stevens Act, there is a national standard for bycatch minimization. Bycatch is officially defined as "fish which are harvested in a fishery, but which are not sold or kept for personal use" (i.e., the dead discards referred to earlier). However, this does not include "fish which are released alive under a recreational catch-and-release fishery management program" (Crowder and Murawski 1998). Of three scenarios these authors describe in their article, I chose to depict the third, entitled "One Man's Trash." This is a situation where, as initial "high-value" species are overfished, fishermen are prompted to diversify their targets, perhaps in combination with emergence of new markets (e.g., shark-fin soup). In the conquest of initially highly valued species, nontargets are at first discarded at a high rate. In the extreme, target species are abandoned for economic reasons, and what used to be bycatch species are now sought. This situation results in allocation conflicts and suboptimal use of both target species and discards (Crowder and Murawski 1998), while at the same time reducing the population's average spawning potential (i.e., a lose-lose situation).

WHO GETS FIRST "DIBS"?

As Wallace et al. (1994) point out, group conflicts arise when one (public owner) perceives that another is getting more. Furthermore,

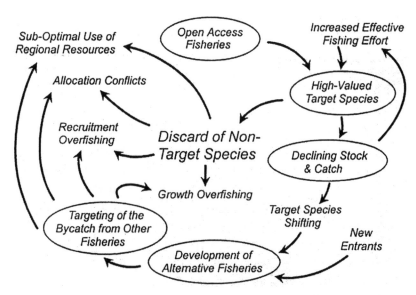

Fig. 5.7. "One Man's Trash," a bycatch scenario. Reprinted courtesy of L. B. Crowder and S. A. Murawski. 1998. Fisheries Bycatch: Implications for Management. *Fisheries* 23 (6): 11.

although management of fishery resources is ultimately the responsibility of elected officials, they delegate that responsibility to federal agencies. In the case of marine fisheries, the agency (NMFS) in turn delegates responsibility for writing FMPs, coordinating meetings, and holding public hearings on proposed rule making to regional fisheries management councils, such as the Mid-Atlantic Fisheries Management Council, comprised of state fisheries managers and other stakeholders. The ASMFC, one of three interstate fisheries commissions (the others being the Gulf and Pacific states commissions), is then charged with making sure the states abide by these FMPs, pursuant to the Atlantic Coastal Fisheries Cooperative Management Act of 1993 (Wallace et al. 1994).

The fisheries biologist is supposed to come up with an estimate of how many fish (or pounds of fish) can be safely harvested without driving the population to its demise, while the council decides how such quotas are allocated. The idea is to manage total fishing mortality, or "total allowable catch" (TAC), in a way that is fair to all participants. Naturally, however, debates over fair allocation usually get pretty heated. In theory, traditional users of the resource, such as bluefin tuna anglers, are supposed to be granted seniority. Historically, though, management decisions have been based on the concept of "maximum sustainable yield" (MSY) from the fishery (a construct rooted in commercial fishing theory).

The Magnuson-Stevens Act, renamed the Sustainable Fisheries Act when it was reauthorized in 1996, incorporates "optimum yield" (OY), a concept stressed decades ago by Stroud (1970), as the first of seven standards for the fishery manager to follow. Optimum yield is based on the maximum amount of fish that can be harvested, modified by socioeconomic and ecological factors. That (downward) adjustment has been invoked in order to allocate harvest in a way that reflects the greatest benefit to the nation, especially with regard to food production and recreational opportunities (Wallace et al. 1994). It is thought that the use of OY should result in more equitable allocations, while hopefully permitting those species whose populations stand at critically low levels to make a comeback, such as the striped bass was able to do (though some aspersions are now being cast on the reality of that comeback).

The entire public process for establishing regulations and allocations for saltwater fishes, in addition to more detail about fisheries management procedures, is spelled out in Wallace et al. (1994), *Fisheries Management for Fishermen: A Manual for Helping Fishermen Understand the Federal Management Process*. This is recommended reading for those interested in getting more involved in the process, and is available from the Auburn University Marine Extension and Research Center in Mobile, Alabama.

Within the overall framework of the NMFS stock determinations and rule making, the states choose how they want to regulate saltwater species within 3 miles, which is why, for example, New Jersey's recently adopted fluke regulations differ from those of New York and Delaware relative to closed seasons, quotas, and/or size limits. Our own Marine Fisheries Council handles that responsibility, whereas our volunteer Fish and Game Council, officially established in 1945, is the key entity responsible for making the rules concerning freshwater fishing seasons, limits, and methods.

CHAPTER 6

Fisheries History and Modern Activities

T oo often, being caught up in day-to-day interests, we tend to ignore history. That's why I've woven the subject throughout this
book. I have been in this field for almost thirty years, following graduate school, and it wasn't until I started researching this chapter that
I uncovered some peculiar laws. Would you believe, for example, that
as recently as 1958 Congress promulgated a Dogfish Shark Eradication Act? The act mandated that the states investigate the distribution and abundance of dogfish, perform experiments to control them,
and initiate a "vigorous program for the elimination and eradication
or development of economic uses of dogfish shark populations." Many
commercial fishermen still maintain that protection of dogfish today
is detrimental to the recovery of groundfish stocks that they prey
upon. Lane and Comac (1992) did not mention this act in their
book *Sharks Don't Get Cancer*, which, based on their chronology of
benchmarks beginning in the early seventies, suggests that medicinal
research pursuant to this act was not actively promoted by the
government.

Fisheries: An Old and Important Profession

Prior to 1871, ten states had established fisheries commissions, starting with Massachusetts in 1865, but the federal government was not
involved in fish or fisheries then (Thompson 1970; McHugh 1970).
In 1871 the American Fish Culturists' Association (AFCA) and the
United States Commission on Fisheries were organized. The latter
organization was the predecessor of the U.S. Bureau of Fisheries, later
to become the U.S. Fish and Wildlife Service. The Commission on

Fisheries began operations on a modest budget at Woods Hole, Massachusetts, in 1871, giving rise to the Woods Hole Oceanographic Institution (WHOI) Marine Biological Laboratory. The second of the commission's labs was created in Beaufort, North Carolina, in 1901 (McHugh 1970). These organizations were, at first, primarily concerned with fish production, but by 1875 the commission had started showing concern over pollution (Thompson 1970).

In 1886 the AFCA was transformed into the American Fisheries Society (AFS), an organization devoted not only to fish culture, but to research concerning both natural and artificially propagated species. The AFS's objective was "to promote the cause of fish-culture; to gather and diffuse information bearing upon its practical success, and upon all matters relating to the interests of fish-culture and the fisheries; and the treatment of all questions regarding fish, of a scientific and economic character" (Thompson 1970). For a long time now, the AFS has been intimately involved in fisheries culture and management topics, and has dispatched a number of influential publications, foremost among them being the *Transactions of the American Fisheries Society*. The AFS certifies fishery scientists, and is organized about a number of geographic "chapters" and professional "interest" sections, including those concerned with marine fisheries, estuarine fisheries, early life-history stages, introduced species, fishery management, water quality, education, genetics, physiology, fish culture and health, international fisheries, and fisheries law. The AFS had about 250 members ca. 1880, 1,000 ca. 1940, 2,500 ca. 1960, 5,000 ca. 1970, and (due to nineties shifts in employment market drivers) 4,000 ca. 1998. Still, the AFS remains a major mover in the business of fisheries science.

The International Council for the Exploration of the Sea (ICES) is credited with being the oldest true international fisheries research organization in the world, having been created in 1902 for the cooperative inquiry into fishery oceanography (McHugh 1970). The Atlantic States Marine Fisheries Compact, which created the Atlantic States Marine Fisheries Commission (ASMFC), was formed in 1942, followed shortly thereafter by passage of the Northwest Atlantic Fisheries Act of 1950. That act created the International Commission for the Northwest Atlantic Fisheries (ICNAF), a body of international scientists that conducts research, maintains landings statistics on high-seas fisheries, and develops management proposals for member countries. In 1949 a consortium of tackle manufacturers got together to create and fund the Sport Fishing Institute (SFI), an organization dedicated to promoting fishery biology and improving sport fisheries management. Through its grants, that organization has gone a long way to enhance fishing opportunities in urban as well as rural environments.

During the first half of the twentieth century, the fishery biologist's tools were fairly simple, like nets, dredges, and water bottle samplers like the one illustrated. Beginning in the fifties, however, the science saw the dawning of much more sophisticated equipment and vessels. Not only were water sampling devices improved and electronic fishfinders added to the biologists' means of bettering their netting efficiency, but as readers have no doubt seen on the "telly," minisubs, unmanned submersible probes, and remote tracking devices were perfected to go places we can't.

The Fish and Wildlife Act of 1956 pulled the Bureau of Commercial Fisheries out of the Department of Commerce, and the Bureau of Sport Fisheries out of the Department of Agriculture, thus establishing the U.S. Fish and Wildlife Service under the direction of the Department of the Interior (DOI). That lasted until 1970, when, by presidential proclamation, the Bureau of Commercial Fisheries was put back under the domain of Commerce once again, now being called the National Marine Fisheries Service (NMFS), which was directly under the auspices of the (also) newly created National Oceanographic and Atmospheric Administration (NOAA). The Bureau of Sport Fisheries remained under the management of the Department of the Interior. Meanwhile, in our area, New Jersey and Delaware had formed the Delaware and New Jersey Compact, thus creating the Delaware River and Bay Authority when that waterbody was sorely in need of help due to pollution.

In terms of habitat and water quality law, 1969 was a blockbuster year. The federal National Environmental Policy Act (NEPA) was promulgated, giving birth to the Environmental Protection Agency (EPA). The NEPA created innumerable career opportunities for fishery biologists and all sorts of other "-ologists" because of its require-

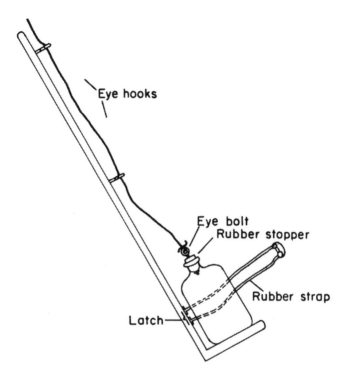

Fig. 6.1. Old-fashioned water bottle sampler. From *Water Quality in Warm-water Fish Ponds* (1979), 208. Reprinted courtesy of Claude Boyd.

ment that every major project, from power plant siting to transmission line routing or harbor dredging, be preceded by submission of an environmental report (ER) by the applicant, and preparation of an environmental impact statement (EIS) by the lead government agency charged with reviewing and determining the merits of the application.

For the fishery biologist, some of the steam went out of the job market in the mid- through late eighties when industrial development slowed and public concern swayed environmental attention more toward hazardous waste site studies and cleanup, which employed relatively greater numbers of geologists, chemists, and engineers. Nevertheless, even cleanup remedies required by order of the Comprehensive Environmental Response, Compensation, and Liability Act (CERCLA, or "Superfund," geared toward defunct operations) or the Resource Conservation and Recovery Act (RCRA, pertaining to active sites) must be protective of organisms and ecosystems. Since Department of Defense facilities are commonly affected by these rules, the army, navy, and air force environmental centers have gotten together to form a "Tri-Service Environmental Support Centers Coordinating Committee" and (among other things) publish guidelines for uniform ecological risk assessments at military installations.

A Chronology of Laws Affecting Fishermen or Fish

Table 6.1 lists the major acts affecting fisheries management and conservation, habitat, and water quality protection. Those not prefaced by "NJ" were all enacted at the federal level, but each state in our area has its own analogue for some (e.g., the Clean Water Act, or CWA). Many of the acts listed have been discussed in previous chapters, so I'm just going to succinctly state the highlights of some of the others whose titles are not necessarily self-explanatory.

The River and Harbors Act, of fundamental importance to dredging matters, authorized the Army Corps of Engineers to regulate "obstructions to navigation." The corps is also granted authority to regulate, given EPA concurrence, dredge, fill, and disposal activities under section 404 of the CWA, and is required, pursuant to the Coastal Zone Management Act, to coordinate permit reviews with the states. The Fish and Wildlife Coordination Act mandates that fish and wildlife conservation have to be given equal consideration and be coordinated with other water resource development projects. The Water Protection and Flood Prevention Act deals with fish and wildlife conservation at small watershed projects having a potential for erosion, sedimentation, and flood damage.

Two acts of interest relative to commercial fishing are the U.S. Fishing Fleet Improvement Act, which created funding for that purpose, and the Commercial Fisheries Research and Development Act of 1964, which authorized the Secretary of the Interior (when commercial fisheries were under the DOI) to give money to the states in the event of a fishery failure due to resource disasters. Some commercial fishermen are now questioning the value of low-interest-loan policies provided to fishing families during the late seventies. At hearings concerning groundfish (cod, flounder, etc.) before the New England Fishery Management Council in 1998, arguments were registered that these loan policies resulted in overcapitalization of the fleet, leading immediately to overfishing and (now) further financial hardship due to increased regulation. The Marine Resources and Engineering Development Act, conversely, broadened the federal fisheries research program with increased emphasis on fundamental research and funding of more labs (like WHOI) and their research vessels (like the *Albatross*).

One of two acts of central importance to sport fishermen is the Black Bass Act of 1926, which prohibited the sale of black bass (e.g., large- and smallmouth bass) on a nationwide basis, totally restricting these basses to recreational use. The other, which has been of pivotal importance in funding state fisheries management and restoration work, is the Dingell-Johnson Act. This funds the states in direct proportion to their sales of fishing licenses, using revenues from excise taxes on sales of all sport fishing equipment. The Federal Aid in Fish Restoration Act has also established a mechanism for providing Wallop-Breaux (W-B) funding for renovation and maintenance of marina pump-out and dumping stations as well as boater education programs, monies coming this time from excise taxes and import duties—at the manufacturer's level—on boats and some kinds of tackle. This is all part and parcel of the Federal Clean Vessel Act (CVA), which makes it illegal to discard plastic and other garbage in coastal waters. More information about the CVA, including a variety of educational materials, can be obtained from the New Jersey Marine Sciences Consortium at Sandy Hook.

Finally, an act special to New Jersey, the Coastal Area Facilities Review Act (CAFRA). The intent of CAFRA was to limit development in the wrong places along the coast, such as wetlands, through a planning, impact assessment, and permitting process. Needless to say, the act has had its share of friends and opponents. The most bedeviling factor in my opinion was the vote in the seventies by citizens of the state itself to legalize gambling in Atlantic City. CAFRA was amended in 1993 to require close coordination with the State

Table 6.1 Major Statutes Affecting Fishes and Their Habitat

River and Harbors Act	1899
Black Bass Act	1926
Fish and Wildlife Coordination Act	1934
Atlantic States Marine Fisheries Compact	1942
Federal Water Pollution Control Act	1948
Federal Aid in Fish Restoration (Dingell-Johnson) Act	1950
Northwest Atlantic Fisheries Act of 1950	1950
Tuna Conventions Act of 1950	1950
Submerged Lands Act	1953
Water Protection and Flood Prevention Act	1954
Fish and Wildlife Act of 1956	1956
Dogfish Shark Eradication Act	1958
U.S. Fishing Fleet Improvement Act	1960
Delaware and New Jersey Compact	1962
Commercial Fisheries Research and Development Act of 1964	1964
Anadromous Fish Conservation Act	1965
Federal Water Quality Act	1965
Marine Resources and Engineering Development Act	1966
National Environmental Policy Act (NEPA)	1969
NJ Wetlands Act	1970

Planning Office, established with passage of the State Planning Act in 1985, to, among other things, identify "coastal centers" for development rather than allow piecemeal encroachment on undisturbed habitat.

Activities You Can Immerse Yourself In

There are two programs I want to promote here. The first, in which Bruce now participates, is a voluntary fish tag-and-release program for inshore saltwater fish run by the American Littoral Society (ALS), whose New Jersey offices are located in Sandy Hook (officially listed as Highlands, N.J.). In this program, volunteers purchase their tag-

210

Federal Water Pollution Control Act	1972
Coastal Zone Management Act	1972
Endangered Species Act	1973
NJ Coastal Area Facilities Review Act (CAFRA)	1973
Atlantic Tunas Convention Act of 1975	1975
Magnuson Fishery Conservation and Management Act	1976
Clean Water Act (CWA)	1977
Fish and Wildlife Improvement Act	1978
Comprehensive Environmental Response, Compensation, and Liability Act (CERCLA)	1980
Atlantic Striped Bass Conservation and Management Act	1984
NJ State Planning Act	1985
Superfund Amendment and Reauthorization Act (SARA)	1987
Water Quality Act	1987
NJ Freshwater Wetlands Protection Act	1988
Federal Clean Vessel Act (CVA)	1992
Water Resources and Development Act of 1992	1992
Atlantic Coastal Fisheries Cooperative Management Act	1993
Sustainable Fisheries Act	1996

ging kits (for a very small price) and are trained in how to handle and tag the fish they catch. Tag-recapture studies are of paramount importance to learning more about fish movements, growth, and mortality rates. As noted in chapter 5, this information is critical to figuring out how to manage the fishery for the benefit of the fish and all those parties who make their living from the sea or simply want to continue to enjoy their favorite form of recreation. Bruce got into it when he caught a fish that had been tagged by another volunteer and returned the tag and details of where and when it was caught to the ALS. He quickly received responses and certificates of appreciation from both the ALS and U.S. Fish and Wildlife Service, and will be listed in the ALS diary of tagging returns. Even if one does not become an active partner in the tagging program, the importance of reporting capture of tagged fish to the ALS and/or NJDEP cannot be stressed enough. The New Jersey Department of Fish, Game and Wildlife (FG&W) also conducts a reef fish tagging effort (report your catch to the FG&W office in Port Republic), and all of the sea-run brown trout have had their adipose fin clipped (a form of mark-recapture study used for the same purposes). Anyone catching one of those fish, of which more than 49,000 had been stocked in the lower Manasquan River as of 1999, is asked to report the catch to the FG&W at the Pequest hatchery in Oxford. Other tag-and-release programs are sponsored by the NMFS in Narragansett, Rhode Island (sharks), and Miami, Florida (tunas); and the Billfish Foundation in Fort Lauderdale, Florida.

In closing, for those of you who would like to experience fisheries fieldwork firsthand, I suggest getting in touch with the FG&W offices and asking about the division's volunteer Wildlife Conservation Corps. It's challenging work, but helping to enhance stream habitat, as was done on Flanders Brook and the Musconetcong River, or lending a hand aboard the NJDEP's saltwater fish survey trawler might be very educational and even spine tingling.

Now, without further ado, I'm going fishing!

Bibliography

Able, Kenneth, and Michael Fahay. 1998. *The First Year in the Life of Estuarine Fishes in the Middle Atlantic Bight.* New Brunswick, N.J.: Rutgers Univ. Press.

Able, Kenneth, and Susan Kaiser. 1994. *New Jersey Estuaries Finfish Resource Assessment Phase I: Literature Summary.* New Brunswick, N.J.: Institute of Marine and Coastal Sciences, Rutgers University.

AFS. 1970. *A List of Common and Scientific Names of Fishes.* American Fisheries Society Special Pub. no. 6. Washington, D.C.

———. 1971. *Stream Channelization: A Symposium.* American Fisheries Society Special Pub. no. 2. Washington, D.C.

———. 1988. *Science, Law and Hudson River Power Plants: A Case Study in Environmental Impact Assessment.* American Fisheries Society Monograph no. 4. Bethesda, Md.

———. 1991. *Common and Scientific Names of Fishes from the United States and Canada.* American Fisheries Society Special Pub. no. 20. Bethesda, Md.

AFS, American Oceans Campaign, and NOAA. 1998. *Protecting and Restoring Essential Fish Habitat.* Bethesda, Md.: American Fisheries Society.

ANSP. 1998. *Aquatic Biomonitoring in the Delaware Water Gap National Recreational Area, Using Benthic Macroinvertebrates: 1995.* Report 98-11. Philadelphia, Pa.: Patrick Center for Environmental Research, Academy of Natural Sciences of Philadelphia.

APHA (American Public Health Assn.), AWWA (American Water Works Assn.), and WPCF (Water Pollution Control Federation). 1980. *Standard Methods for the Examination of Water and Wastewater.* 15th ed. Washington, D.C.: APHA.

Armantrout, Neil. 1998. *Aquatic Habitat Inventory Terminology Glossary.* Bethesda, Md.: Western Div., American Fisheries Society.

Barnes, Robert D. 1963. *Invertebrate Zoology.* Philadelphia, Pa.: W. B. Saunders Co.

Barrett, Pete. 1992. *Fishing for Tuna and Marlin*. Point Pleasant, N.J.: Fisherman Library Corp.

Beck, Scott. 1995. "White Perch." In *Living Resources of the Delaware Estuary*, ed. Louise Dove and Robert Nyman, 235–243. Delaware Estuary Program, U.S. Environmental Protection Agency.

Bergman, Raymond. 1938. *Trout*. New York: Alfred A. Knopf.

Bigelow, Henry, and William Schroeder. 1953. *Fishes of the Gulf of Maine*. Fishery Bulletin 74, vol. 53, U.S. Fish and Wildlife Service. Washington, D.C.

Boa, Marty. 1994. "Yellow Brook/Monmouth County's Hidden Trout Stream." In *Discovering and Exploring New Jersey's Fishing Streams and the Delaware River*, ed. Steve Perrone, 100–101. Somerdale, N.J.: New Jersey Sportsmen's Guides.

Boyd, Claude. 1979. *Water Quality in Warmwater Fish Ponds*. Auburn, Ala.: Auburn University Agricultural Experiment Station.

Boyer, Michael. 1995. "Carp" and "Minnows." In *Living Resources of the Delaware Estuary*, ed. Louise Dove and Robert Nyman, 151–156, 227–233. Delaware Estuary Program, U.S. Environmental Protection Agency.

Bryant, T. L., and J. R. Pennock. 1988. *The Delaware Estuary: Rediscovering a Forgotten Resource*. Newark, Del.: Univ. of Delaware Sea Grant College.

Bulloch, David. 1986. *Marine Gamefish of the Middle Atlantic*. Highlands, N.J.: American Littoral Society.

Camhi, Merry. 1998. *Sharks on the Line*. Islip, N.Y.: National Audubon Society.

Caputi, Gary. 1993. *Fishing for Striped Bass*. Point Pleasant, N.J.: Fisherman Library Corp.

Carlander, Kenneth. 1969. *Handbook of Freshwater Fishery Biology*. Vol. 1. Ames, Iowa: Iowa State Univ. Press.

Caucci, Al, and Bob Nastasi. 1984. *Instant Mayfly Identification Guide*. Henryville, Pa.: Comparahatch.

Cawley, James, and Margaret Cawley. 1971. *Exploring the Little Rivers of New Jersey*. New Brunswick, N.J.: Rutgers Univ. Press.

Christy, Francis, Jr., and Anthony Scott. 1965. *The Common Wealth in Ocean Fisheries*. Baltimore, Md.: Johns Hopkins Univ. Press.

Coble, Daniel. 1975. "Smallmouth Bass." In *Black Bass Biology and Management*, 21–33. Washington, D.C.: Sport Fishing Institute.

Coker, R. F. 1954. *This Great and Wide Sea*. New York: Harper Torchbook.

Crowder, Larry, and Steven Murawski. 1998. "Fisheries Bycatch: Implications for Management." *Fisheries* (American Fisheries Society, Bethesda, Md.) 23 (6): 8–17.

Cunningham, John. 1958. *The New Jersey Shore*. New Brunswick, N.J.: Rutgers Univ. Press.

Daiber, Franklin. 1995. "Sharks, Skates, and Rays." In *Living Resources of the Delaware Estuary*, ed. Louise Dove and Robert Nyman, 259–265. Delaware Estuary Program, U.S. Environmental Protection Agency.

Daignault, Frank. 1996. *Striper Hot Spots: The 100 Top Surf Fishing Locations from New Jersey to Maine*. Old Saybrook, Conn.: Globe Pequot Press.

Dix, Charles. 1995. "Yellow Perch." In *Living Resources of the Delaware Estuary*, ed. Louise Dove and Robert Nyman, 245–249. Delaware Estuary Program, U.S. Environmental Protection Agency.

Eastby, Allen. 1994a. "The Paulinskill—Three Faces of the River." In *Discovering and Exploring New Jersey's Fishing Streams and the Delaware River*, ed. Steve Perrone, 46–49. Somerdale, N.J.: New Jersey Sportsmen's Guides.

———. 1994b. "The Pequest—Easy Access." In *Discovering and Exploring New Jersey's Fishing Streams and the Delaware River*, ed. Steve Perrone, 52–54. Somerdale, N.J.: New Jersey Sportsmen's Guides.

Eddy, Samuel, and James Underhill. 1974. *Northern Fishes*. Minneapolis, Minn.: Univ. of Minnesota Press.

Eipper, A. W., and H. A. Regier. N.d. *Fish Management in New York Farm Ponds*. New York State College of Agriculture and Life Sciences Extension Bulletin 1089. Ithaca, N.Y.: Cornell University.

Fahay, Michael. 1995. "American Eel." In *Living Resources of the Delaware Estuary*, ed. Louise Dove and Robert Nyman, 175–181. Delaware Estuary Program, U.S. Environmental Protection Agency.

Figley, W. K. 1984. *New Jersey's Recreational Ocean Shark Fishery—1984*. Marine Fisheries Administration Information Series 84-1. Trenton, N.J. Cited by Camhi 1998.

Fisherman Press. 1993. *The Guide to Blackfish and Sea Bass*. Point Pleasant, N.J.: Fisherman Library Corp.

Geisel, Theodore, and Audrey Geisel. 1971. "The Lorax." In *Six by Seuss*, Theodore and Audrey Geisel. 1991. New York: Random House.

Geiser, John. 1969. "Shore Area Fish." Series of daily articles appearing in the former *Asbury Park Evening Press* from June 1969 through about October 1969. Asbury Park, N.J.

Hall, William. 1995. "Atlantic Menhaden." In *Living Resources of the Delaware Estuary*, ed. Louise Dove and Robert Nyman, 219–225. Delaware Estuary Program, U.S. Environmental Protection Agency.

Himchak, Peter. 1998. "Tautog—A Profile." *Reef News*. Port Republic, N.J.: N.J. Dept. of Environmental Protection.

Hynes, H. B. N. 1970. *The Ecology of Running Waters*. Toronto, Canada: Univ. of Toronto Press.

Johns, Fred. 1994a. "Sandts Eddy Shad" and "Toll Bridge Stripers." In *Discovering and Exploring New Jersey's Fishing Streams and the Delaware River*, ed. Steve Perrone, 120. Somerdale, N.J.: New Jersey Sportsmen's Guides.

———. 1994b. "Toll Bridge Stripers." In *Discovering and Exploring New Jersey's Fishing Streams and the Delaware River*, ed. Steve Perrone, 142–143. Somerdale, N.J.: New Jersey Sportsmen's Guides.

Kamienski, Don. 1987. *Fluke Fishing* (video). Point Pleasant, N.J.: Fisherman Library Corp.

———. 1993. *Fishing for Fluke*. Point Pleasant, N.J.: Fisherman Library Corp.

Lager, J. A., W. G. Smith, W. G. Lynard, R. M. Finn, and E. J. Finnemore. 1977. *Urban Stormwater Management and Technology: Update and User's Guide*. EPA-600/8-77-014. Cincinnati, Ohio: U.S. Environmental Protection Agency.

Lagler, Karl. 1956. *Freshwater Fishery Biology*. Dubuque, Iowa: Wm. C. Brown Co.

Lagler, Karl, John Bardach, and Robert Miller. 1962. *Ichthyology*. New York, N.Y.: John Wiley & Sons.

Lane, William, and Linda Comac. 1992. *Sharks Don't Get Cancer*. Garden City Park, N.Y.: Avery Publishing Group.

Lanzim, Marc. 1994. "The Toms River: Route 70 to County Road 571." In *Discovering and Exploring New Jersey's Fishing Streams and the Delaware River*, ed. Steve Perrone, 65–66. Somerdale, N.J.: New Jersey Sportsmen's Guides.

Laycock, George. 1966. *The Alien Animals*. Garden City, N.Y.: Natural History Press.

Lippson, Alice, and Robert Lippson. 1997. *Life in the Chesapeake Bay*. Baltimore, Md.: Johns Hopkins Univ. Press.

LMS (Lawler, Matusky & Skelly, Engineers). 1995. *Abundance and Stock Characteristics of the Atlantic Tomcod Spawning Population in the Hudson River, Winter 1994–1995*. New York Power Authority.

Long, John. 1995. *The Rise of Fishes*. Baltimore, Md.: Johns Hopkins Univ. Press.

Lorenzetti, Al. 1995. *Fluke Fishing—Improving Your Catch* (video). Point Pleasant, N.J.: Fisherman Library Corp.

———. 1995. *Live Bait Striped Bass* (video). Point Pleasant, N.J.: Fisherman Library Corp.

———. 1996. *Fishing for Bluefish* (video). Point Pleasant, N.J.: Fisherman Library Corp.

Luftglass, Manny. 1996. *Gone Fishin' for Carp!* Self-published.

Luftglass, Manny, and Ron Bern. 1998. *Gone Fishin': The 100 Best Spots in New Jersey.* New Brunswick, N.J.: Rutgers Univ. Press.

MacCrimmon, Hugh, and William Robbins. 1975. "Distribution of the Black Basses in North America." In *Black Bass Biology and Management,* 56–66. Washington, D.C.: Sport Fishing Institute.

Maharaj, V., and J. E. Carpenter. 1997. *The 1996 Economic Impact of Sport Fishing in New Jersey.* Alexandria, Va.: American Sportfishing Assn. Cited by Camhi 1998.

Malat, Joe. 1993. *Surf Fishing.* Nags Head, N.C., and York, Pa.: Wellspring.

McBride, Richard. 1995. "Marine Forage Fish." In *Living Resources of the Delaware Estuary,* ed. Louise Dove and Robert Nyman, 211–217. West Trenton, N.J.: Delaware Estuary Program, U.S. Environmental Protection Agency.

McFadden, James. 1977. *Influence of Indian Point Unit 2 and Other Steam Electric Plants on the Hudson River Estuary, with Emphasis on Striped Bass and Other Fish Populations.* Report to the Consolidated Edison Company, New York, N.Y.

McHugh, J. L. 1970. "Trends in Fishery Research." In *A Century of Fisheries in North America,* ed. N. Benson, 25–56. American Fisheries Society Special Pub. no. 7. Washington, D.C.

McLaren, J. B., J. C. Cooper, T. B. Hoff, and V. Lander. 1981. "Movements of Hudson River Striped Bass." *Trans. Amer. Fish. Soc.* 110: 158–167.

Metcalf, Greg. 1994. *Fishing for Giant Bluefin Tuna* (video). Point Pleasant, N.J.: Fisherman Library Corp.

Methot, Rick. 1994. "Lambertville Shad." In *Discovering and Exploring New Jersey's Fishing Streams and the Delaware River,* ed. Steve Perrone, 114–117. Somerdale, N.J.: New Jersey Sportsmen's Guides.

Michels, Stewart. 1995. "Drum." In *Living Resources of the Delaware Estuary,* ed. Louise Dove and Robert Nyman, 167–173. West Trenton, N.J.: Delaware Estuary Program, U.S. Environmental Protection Agency.

Miller, Joseph. 1995. "American Shad." In *Living Resources of the Delaware Estuary,* ed. Louise Dove and Robert Nyman, 251–257. West Trenton, N.J.: Delaware Estuary Program, U.S. Environmental Protection Agency.

Miller, Joseph, and A. Lupine. 1996. *Creel Survey of the Delaware River American Shad Recreational Fishery.* Hellertown, Pa.: Delaware River Shad Fisherman's Association.

Needham, James, and Paul Needham. 1962. *A Guide to the Study of Fresh-Water Biology.* San Francisco, Calif.: Holden-Day.

New Jersey Sea Grant. 1994. "Catch and Release Fishing for Snapper Blues." Barnegat Bay Fact Sheet No. 1. Fort Hancock, N.J.: New Jersey Sea Grant Education and Outreach Program.

NJDCD. 1940. *The Geology of New Jersey*. N.J. Dept. of Conservation and Development Bulletin 50, Geologic Series. Trenton, N.J.

NJDEP. 1982. *New Jersey's Recreational and Commercial Ocean Fishing Grounds*. Technical Series 82-1. Trenton, N.J.: N.J. Dept. of Environmental Protection, Marine Fisheries Admin.

————. 1994. *Places to Fish—List of NJ Lakes, Ponds, Reservoirs, and Streams Open to Public Angling*. Trenton, N.J.: N.J. Dept. of Environmental Protection, Bureau of Freshwater Fisheries.

————. N.d. (ca. 1994). *Salt Water Fishing in New Jersey*. Trenton, N.J.: NJ Dept. of Environmental Protection, Marine Fisheries Admin.

————. N.d. (ca. 1994). *New Jersey Party and Charter Boat Directory*. Trenton, N.J.: N.J. Dept. of Environmental Protection, Marine Fisheries Admin.

————. 1998a. *New Jersey 1996 State Water Quality Inventory Report*. Trenton, N.J.: N.J. Dept. of Environmental Protection, Office of Environmental Planning.

————. 1998b. *New Jersey Fish and Wildlife Digest*. Trenton, N.J.: N.J. Dept. of Environmental Protection, Div. of Fish, Game and Wildlife.

————. 1998c. *Inventory of N.J. Lakes and Ponds*. Trenton, N.J.: N.J. Dept. of Environmental Protection, Div. of Fish, Game and Wildlife.

————. 1998d. *List of Warmwater and Coolwater Sportfish Reared and Stocked by the New Jersey Division of Fish, Game and Wildlife during the 1990–1997 Period*. Hackettstown, N.J.: NJ Dept. of Environmental Protection, Bureau of Freshwater Fisheries.

————. 1998e. *NJ Reef News: 1998 Annual Edition*. Trenton, N.J.: N.J. Dept. of Environmental Protection, Div. of Fish, Game and Wildlife.

NJMR. 1996. *The Business of Dredging*. Trenton, N.J.: New Jersey Maritime Resources, N.J. Dept. of Commerce and Economic Development.

Odum, Eugene. 1959. *Fundamentals of Ecology*. Philadelphia, Pa.: W. B. Saunders Co.

Palisades Interstate Park Commission. 1998. *Notes* 2 (no. 1).

Peinecke, Al. 1994a. "The Big Flatbrook." In *Discovering and Exploring New Jersey's Fishing Streams and the Delaware River*, Steve Perrone, 12–13. Somerdale, N.J.: New Jersey Sportsmen's Guides.

————. 1994b. "Dark Moon Brook/Bear Creek." In *Discovering and Exploring New Jersey's Fishing Streams and the Delaware River*, Steve Perrone, 72–73. Somerdale, N.J.: New Jersey Sportsmen's Guides.

218

Perrone, Steve. ed. 1994a. *Discovering and Exploring New Jersey's Fishing Streams and the Delaware River*. Somerdale, N.J.: New Jersey Sportsmen's Guides.

———. 1994b. *New Jersey Lake Survey Fishing Maps Guide*. Somerdale, N.J.: New Jersey Sportsmen's Guides.

Piehler, Glenn. 1972. "An Investigation of the Massachusetts Marine Sport Fisheries Including a Critique of a Probability Proportional to Size Sampling Method of Estimating Numbers and Attributes of Salt Water Anglers." Ph.D. dissertation, University of Massachusetts.

Pratt, H. L., and R. R. Merson. 1997. *Sandbar Shark Nursery Grounds Project: Report of the 1996 Apex Predators Investigation*. Narragansett, R.I.: Natl. Oceanographic and Atmospheric Admin. and Natl. Marine Fisheries Service. Cited by Camhi 1998.

Quinn, John. 1997. *Fields of Sun and Grass: An Artist's Journal of the New Jersey Meadowlands*. New Brunswick, N.J.: Rutgers Univ. Press.

Reid, George. 1961. *Ecology of Inland Waters and Estuaries*. New York: Reinhold Publishing Corp.

———. 1967. *Pond Life*. New York: Golden Press.

Reschke, Carol. 1990. *Ecological Communities of New York State*. Latham, N.Y.: New York State Department of Environmental Conservation.

Richey, David. 1980. *The Fly Hatches*. New York: Hawthorn Books.

Ristori, Al. 1995. *Fishing for Bluefish*. Point Pleasant, N.J.: Fisherman Library Corp.

Rounsefell, George, and W. Harry Everhart. 1953. *Fishery Science: Its Methods and Applications*. New York: John Wiley & Sons.

Rowe, Barney. 1993. *Catching More Fresh Water Fish*. Ocean City, Md., and York, Pa.: Wellspring.

Rowsome, Frank Jr. 1965. *The Verse by the Side of the Road*. New York: Penguin Group.

Ruppel, Bruce. 1994. "Delaware River Muskellung Fishing." In *Discovering and Exploring New Jersey's Fishing Streams and the Delaware River*, ed. Steve Perrone, 126–133. Somerdale, N.J.: New Jersey Sportsmen's Guides.

Ruttner, Franz. 1963. *Fundamentals of Limnology*. Toronto, Canada: Univ. of Toronto Press.

Scholl, Dennis. 1994. "24 Miles of Delaware River Shad Fishing." In *Discovering and Exploring New Jersey's Fishing Streams and the Delaware River*, ed. Steve Perrone, 106–113. Somerdale, N.J.: New Jersey Sportsmen's Guides.

Schwiebert, Ernest, Jr. 1955. *Matching the Hatch*. New York: Macmillan Co.

Seagraves, Richard. 1995. "Weakfish." In *Living Resources of the Delaware*

Estuary, ed. Louise Dove and Robert Nyman, 293–298. West Trenton, N.J.: Delaware Estuary Program, U.S. Environmental Protection Agency.

Smith, Kelly. 1995. "Brackish Water Killifish." In *Living Resources of the Delaware Estuary*, ed. Louise Dove and Robert Nyman, 199–210. West Trenton, N.J.: Delaware Estuary Program, U.S. Environmental Protection Agency.

Stanne, Stephen, Roger Panetta, and Brian Forist. 1996. *The Hudson: An Illustrated Guide to the Living River*. New Brunswick, N.J.: Rutgers Univ. Press.

Steimle, Frank. 1995. "Structure-Oriented (Reef) Fish." In *Living Resources of the Delaware Estuary*, ed. Louise Dove and Robert Nyman, 267–273. West Trenton, N.J.: Delaware Estuary Program, U.S. Environmental Protection Agency.

Stoughton, Candace. 1997. Health and Ecological Criteria Div., USEPA, internet correspondence, http://h2O.usgs.gov/public/wicp/appendixes/ AppendG., March 17.

Stroud, Richard. 1970. "Future of Fisheries Management in North America." In *A Century of Fisheries in North America*, ed. N. Benson, 291–308. American Fisheries Society Special Pub. no. 7. Washington, D.C.

Thompson, Paul. 1970. "The First Fifty Years—The Exciting Ones." In *A Century of Fisheries in North America*, ed. N. Benson, 1–12. American Fisheries Society Special Pub. no. 7. Washington, D.C.

Trout Unlimited. 1975. *New Jersey Trout Guide*. Edison, N.J.: New Jersey Council, Trout Unlimited.

USEPA. 1991. *Technical Support Document for Water Quality–based Toxics Control*. EPA/505/2-90-001. Washington, D.C.: U.S. Environmental Protection Agency.

Venturo, Greg. 1995. *How to Fish Wrecks, Lumps and Rock Piles*. Point Pleasant, N.J.: Fisherman Library Corp.

Wallace, Richard K., William Hosking, and Stephen T. Szedlmayer. 1994. *Fisheries Management for Fishermen: A Manual for Helping Fishermen Understand the Federal Management Process*. MASGP-94-012. Mobile, Ala.: Auburn University Marine Extension and Research Center.

Walton, Izaak. (1653) 1936. *The Compleat Angler; or, The Contemplative Man's Recreation*. New York: Heritage Press.

Whitmore, W. H. 1997. *Marine Recreational Fishing in Delaware, 1995–1996*. Dover, Del.: Dept. of Natural Resources and Environmental Control. Cited by Camhi 1998.

Wilde, Gene. 1998. "Tournament-associated Mortality in Black Bass." *Fisheries* (American Fisheries Society, Bethesda, Md.) 23 (10): 12–22.

Woodhouse, W. W., E. D. Seneca, and S. W. Broome. 1976. *Propagation and Use of* Spartina alterniflora *for Shoreline Erosion Abatement.* U.S. Army Corps of Engineers Tech. Report no. 76-2. Fort Belvoir, Va.

Zich, H. 1977. *The Collection of Existing Information and Field Investigation of Anadromous Clupeid Spawning in New Jersey.* N.J. Department of Environmental Protection Misc. Report no. 41. Lebanon, N.J.

Index

[Note: italicized page numbers denote figures or illustrations]

About the Author

Born in 1943 in Elizabeth and raised in Union, New Jersey, Glenn Piehler spent his formative summers in Ocean Beach Unit #1, where he was first introduced to fishing. He graduated from Bloomfield College in 1965 and earned a master's in fishery biology at Michigan State University and a doctoral degree from the University of Massachusetts. Since then, he has lived (chronologically) in New Hampshire; Bricktown, Fair Haven, and Highland Lakes, New Jersey; San Francisco; and back again in New Jersey—first Princeton and ultimately Weehawken. Glenn Piehler is also a former guitar teacher, youth league baseball and soccer coach, parent teacher and author of an instructional workbook on pond study for the Fair Haven sixth-grade Stokes Forest retreat, member of Fair Haven's Waterfront Committee and vice president of their Recreation Commission, Webelos Scout leader, and winter Special Olympics skiing volunteer. He now maintains a consulting practice in Hoboken.